The Engine Driver's Manual

The Engine Driver's Manual

How to prepare, fire and drive a steam locomotive

Brian Topping

crecy.co.uk

First published 1998
Reprinted 2000, 2002, 2004, 2005

ISBN 9781800353176

Printed in Turkey by Pelikan

Crécy Publishing Ltd
1a Ringway Trading Estate, Shadowmoss Rd,
Manchester M22 5LH
www.crecy.co.uk

Front cover: On the same morning, Saturday 4 November 1967, an unidentified 8F reverses off shed, having been given priority over a Black 5 on the left. The crews, who will have to take extra care in these conditions, exchange greetings as the 8F disappears into the mist. At this late stage, it seemed that Springs Branch was relying on Black 5s or 8Fs to work all its rosters, although 76075, 76081 and the now-preserved 76077 were still there, maybe either withdrawn or stored. *Allan Heyes*

Back cover: The footplate of 'Schools' class 4-4-0 No 30910 Merchant Taylors recorded at Faversham in 1959. *Marcus Eavis/Online Transport Archive*

MIX
Label | Supporting responsible forestry
FSC® C151955

Contents

Introduction

Ever since railways came into being in the early 19th century it has been the ambition of many small boys and some girls to become the driver of a steam train. In the early years driving a steam locomotive was the equivalent of piloting an aircraft today.

Initially there was a belief that anyone who travelled faster than a galloping horse would suffocate due to lack of air. The railway was at that time the fastest means of travel known to man, and even now, with high speed trains throughout the world such as the French TGV *(Train a Grand Vitesse),* it is the fastest method of transport on land available to the general public. Indeed, the drivers of these high-speed trains, which cruise at around 300 kilometres per hour (kph) or 180mph, are known as 'train pilots' by the French railway SNCF.

However, even in these modern high-speed times there are still many people who drive steam engines as volunteers on a regular basis on private railways throughout the country. There are also many other members of the general public whose ambition is to drive steam locomotives, and this is made possible by the driver training courses operated at many railway centres throughout the country.

I was lucky to start my railway career before the end of steam in the late 1950s at Bury Motive Power Depot, where I began as an engine cleaner, rising through promotion to 'passed cleaner' (a cleaner passed to act as a fireman). Later, during 1961, there was a vacancy for a regular fireman at nearby Rochdale, a subsidiary depot of Bury, and my seniority allowed me to apply for it. The transfer was agreed and I moved as a booked fireman to Rochdale. In 1962, on the advice of my driver, I decided to attended the Mutual Improvement Classes (MICs) held at the University of Manchester Institute of Science & Technology (UMIST) to prepare myself for the two-day drivers' exam that I would be required by the railway to undergo in the near future.

The MICs were held at most motive power depots in the heyday of the railway, and were usually presided over by a senior driver or fireman. Attendance by pupils was on a purely voluntary basis and was unpaid. The subjects taught covered the rules of the railway, as well as the design and construction of the steam locomotive. To assist the pupils in their study the railway used to provide a converted coach, which could be taken to various motive power depots and left for some weeks so that, despite their irregular times of duty, all the students would have a chance to attend. The coach contained various cross-section mock-ups of locomotive parts, some supplied by the railway contractors that manufactured them, which could be used to demonstrate how a steam engine worked. There was also a coach that toured the classes covering railway rules and signalling practice.

In later years the railway held these classes at central points such as UMIST, which at that time served the Manchester area. By then the weekly classes were by no means well attended, with only about ten pupils, and this would be a conservative estimate as two drivers per week would be undergoing the drivers' exam.

At a subsidiary depot such as Rochdale promotion to 'passed fireman' (a fireman that was passed to drive a steam locomotive) was not dependent on the candidate's service or seniority, as was the case at the parent or a larger motive power depot. Thus the chance to attend these classes to assist my understanding of the rules and other railway practice was not to be missed, and served to compensate for my lack of experience. In November 1964, at the age of 23, I took and passed the two-day drivers' exam with the Loco Inspector, and became a passed fireman, in which capacity I served until the end of my railway service. This came about due to the closure of many of the local branch lines and the increase of dieselisation, which caused me to be made redundant with the end of steam in 1968. It would have been possible to transfer to the newer motive power, but for personal reasons I chose not to take that avenue.

My interest in railways did not die, however, but although I still read plenty of books on the subject, I was not able to get involved with any of the embryo railway preservation societies that were opening in the British Isles due to my new occupation as a lorry driver, employed away from home in the British Isles and later on Continental journeys.

In 1983 I was made redundant again, and this time I found employment nearer home, which gave me more time to pursue my interest in steam railways. In 1984 I became involved in the rebuilding of 'Black Five' 4-6-0 o 45337 on the East Lancashire Railway at Bury, and now once again I have the pleasure of being rostered to drive steam-hauled trains on that railway. The excitement and the feel of driving a steam locomotive is hard to put into words, and can only be described by the phrase 'live steam'. Whether the loco is the mighty *Flying Scotsman* or a lowly 'Thomas the Tank Engine', the feeling is fantastic! To be at the controls of a locomotive, working it correctly along the line and bringing it and its train to the end of a journeygives one the special pride of a job well done.

Throughout this book I will try to pass on some of the knowledge I have gained of the various locomotives both on British Railways and on the East Lancashire Railway for the enthusiast who, like myself, aspires to be a driver on one of the many excellent preserved railways throughout the country. It is also hoped that the book will be of interest to the layman who wants to know more about the duties of railwaymen and how a steam locomotive is operated safely and efficiently.

As the career of a railwayman starts at the bottom of the ladder in his chosen department, I will start this book in the same place. Just as many Station Masters started by sweeping the platform, so engine drivers started by cleaning locomotives. This provided a basic knowledge of, and the opportunity to study, the steam locomotive at close quarters. I have taken part in footplate experience training so I make no apology for this approach; indeed, some enthusiasts may have missed some important piece of necessary information not previously gained or omitted, which could lead to a serious problem later in their future career on the footplate. For example, on countless occasions participants on these courses mix up the vacuum ejector with the water injector and their different uses.

I will use the Stanier 'Black Five' as my example locomotive as it is a design that has much in common with other engines used in preservation today, such as the Stanier 8F 2-8-0 freight loco, the 'Jubilee' 4-6-0, the Stanier Class 4 tank engine and the Ivatt Class 2 locomotive. Also, during its life the 'Black Five' was used by British Railways as a test-bed for the future development of the steam locomotive.

Acknowledgements

I would like to acknowledge the assistance of the following with photographs, drawings and technical information: British Railways Board, for their permission to reprint the general drawings of the steam locomotive; Davies & Metcalfe of Romiley, Cheshire, for drawings and information on miectors; the Reference Library of the National Railway Museum, York, for assistance with research and photographs; the high-quality special photographic work carried out by photographers Dave Dyson, Brian Dobbs and Richard Fox, together with those individually credited who have loaned prints; and the East Lancashire Railway and Riley & Son Ltd of Bury, for assistance with the photography of locomotive parts.

I also wish to thank Harry Friend, retired BR Footplate Inspector, North Eastern Region, Roger Johnstone of the East Lancashire Railway, and Mr E. Cave, a former BR driver, for reading and checking the text; Peter Nicholson; and editor Will Adams.

I would like to dedicate this book to the memory of my two German shepherds, Sheba and Thombi, for the hours that they sat by me waiting to go for walks while I worked at the word processor, and my partner Linda, for the encouragement given when I was stuck for an explanation of a locomotive part.

'Crab' 2-6-0 No 42700 leaves Bury Bolton Street station with a special Saturday train to the Lancashire coast. About 400 yards down the line is Bury Motive Power Depot where the author started his railway service in the late 1950s. *Eric Bentley*

1
Anatomy of the steam locomotive

FRAMES AND WHEELS

Mainframes

The main frames of a locomotive are usually inside the wheels, although some older types of locomotive also had frames outboard of the driving wheels; these were known as double-framed locomotives. The frames are built from steel plates about 1-inch-thick, their depth being dependent on the design of the locomotive. Certain engines, mostly of American or Continental design, have what are known as 'bar frames', which are assembled from steel sections.

The frame members are held together at each end by the buffer beams, with additional strength provided by plates set either vertically or horizontally along the length of the structure and known as frame stretchers. It was originally necessary to rivet together the frames, buffer beams and stretchers, as early forms of welding did not allow any flexibility in the assembly.

As the locomotive grew in size the additional length of the rigid frame required to accommodate the larger boiler meant that a design had to be developed whereby the wheels did not only carry the weight but would also be able to pivot within this frame in order to negotiate curves in the rails, of which more in a moment.

Part of the main frame of a standard locomotive showing the horizontal frame stretchers. As can be seen, in this particular example the stretchers run most of the distance below the boiler. There are also additional vertical stretchers. They are made from steel plate, and the triangular pieces cut out allow flexibility as well as keeping the axle weight down. *R. G. Fox*

Wheels and tyres

Locomotive wheels and tyres are made from cast steel or, in some cases on early locomotives, cast iron with steel tyres. During the war years cast iron was again utilised instead of steel, a necessary alternative brought about by the shortage of materials at that time.

As all the power of the locomotive is transmitted to the track through the tyre, it must be firmly fixed to the wheel centre. A tyre made from bestquality rolled steel with no join is fitted over the rim of the wheel by the expansion method. The bore of the tyre is fractionally smaller than the outside diameter of the wheel centre. The tyre is heated and fitted over the rim so that when it cools it shrinks to grip the wheel tightly.

In the case of a driving wheel, where friction caused by slipping or the application of the brakes may create sufficient

A tyre being shrunk on to a wheel. It has been heated by gas jets, lowered on to the wheel centre then allowed to cool; as it contracts it becomes a tight fit on the outer edge of the wheel. *Courtesy ABB Crewe*

heat to expand the tyre, it is also retained on the wheel centre by one of four means:

1 Studs that screw through the rim of the wheel into the tyre. This type of fixing can be found on earlier locomotives.
2 A Gibson ring, like a large circlip that fits in a groove in the inner edge of the tyre.
3 Rivets, usually found on wheels with cast iron centres. Except for some small industrial locomotives, this method went out of fashion very many years ago.
4 A double-lip ring, the outer edge of which has a deep, lipped wall that abuts the wheel centre. The inner edge has a turned lip that is just a little smaller than the bore of the tyre. When the tyre is fitted it is expanded sufficiently to allow the wheel centre to pass through the lip and up to the outer edge of the tyre. As the tyre contracts the wheel centre becomes completely encased by the tyre.

The tyre is about 2 inches thick and serves two very important purposes: the first is to provide strength and durability and the second is to facilitate replacement. The tyre is the only

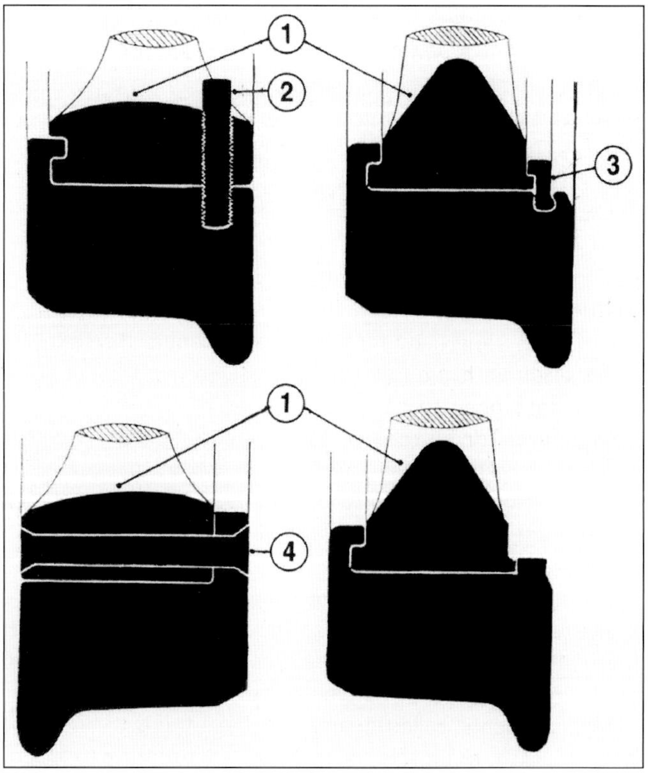

Different ways of fixing tyres to locomotive wheels. Clockwise from top left: studs; the Gibson ring; the double-lip joint; and rivets: 1. wheel centre; 2 stud; 3 Gibson ring; 4 rivet.

The driving wheel of a 'Black Five' locomotive showing the balance weight that compensates for the weight of the big end, crank pin and radius arm. *D. Dyson*

A flangeless driving wheel as fitted to an industrial locomotive. Note that the centre coupling rod has a joint that allows the leading wheels to move in the axlebox when the loco negotiates tight curves. *Author*

part of the wheel that suffers from wear and tear caused by contact with the track and the friction of the brakes. Wear can also be caused by friction during slipping through lack of adhesion with the rail. When tyres are worn to a certain degree they need to be replaced, and it is only a relatively simple operation to remove them and replace them with a new set, thus saving the expense of a complete new wheel.

The wheels are attached to the axle by press fitting at about 60 tons' pressure, and the assembly is then referred to as a wheelset. (The wheels do not rotate on the axle independently of each other.) In addition to the press fit, the driving wheels are keyed on to the axle to keep the crank pins in the correct position. The crank pins at opposite ends of the driven axles on a two-cylinder locomotive are set at 90 degrees to one another, which is called quartering. This is done so that the two pistons do not reach the end of a stroke at the same time, which would cause a dead spot when the locomotive would be unable to move away from a stand. With quartering, when one piston is at the end of the stroke (the position of least effort), the other is halfway down the stroke (the position of maximum effort). It is usual for the crank on the right-hand side of the locomotive to lead the crank pin on the left-hand side by the quarter-turn; this gives four power strokes per revolution of the driving wheels.

The crank pins on a three-cylinder locomotive are set at 120 degrees in order to give six equal power strokes per revolution of the driving wheels.

To compensate for the weight of the crank pins and the connecting and coupling rods of the locomotive, the driving wheels are fitted with counter-balance weights. Those on the wheels that carry the connecting rod crank pins are larger than those on the wheels that only carry the coupling rods, to compensate for the additional weight. If the driving wheels were not balanced, the locomotive would rock from side to side at speed, caused by the oscillating weight of the motion.

Although a type of partial balancing was employed on the ex WD freight engines, for example, it was not enough to compensate for the oscillating weights, and at speed the riding of these locomotives was seriously impaired, with the engine and tender oscillating together. Several Stanier SF Class 2-8-0 freight locomotives had a star painted under the number on the cab side indicating that the wheels had been finely balanced, although these locomotives were not known in particular for bad riding.

As the wheels are fixed securely to the axle there is a need to devise some form of differential to compensate for the varying speed of rotation when the locomotive is negotiating a curve. This is achieved by profile-turning the tyres so that the outer edge of the wheel is slightly smaller than the inner edge by the flange that keeps the locomotive on the rails. When the locomotive is negotiating a turn the engine tends to lean outwards, thus causing the wheelsets to slide within the flange limits between the rails. This action means that the wheels on the inside of the curve run on the smaller-diameter outer edges and those on the outside on the larger diameter inner edges against the flanges. On a sharp curve this is assisted by banking the track slightly to make the locomotive lean inwards; this also helps to steer the locomotive through the curve.

The British Railways Standard 9F class 2-10-0 heavy freight locomotives, together with other ten coupled and some shunting engines, have no flanges on the centre driving wheelset. By this means the wheelset can actually cross the running rail without restriction by a flange, enabling the engine to negotiate the small-radius curves often found in goods yards. The total length of the wheelbase of the '9F' over the five driving wheel centres held in the rigid frame of the locomotive is 21ft Sin, with the total wheelbase of the engine and tender being 56ft 11in.

Axles and axleboxes

A locomotive's axles rotate in axleboxes, which are usually made of cast steel or bronze and move up and down vertically between polished steel slides known as horn guides, fitted to the main frames or bogie of the locomotive or tender. Across the bottom of the horn guide is a heavy forged steel bar, the horn stay, which keeps the axlebox in place and prevents the guide from spreading at the bottom due to the weight of the locomotive.

The vertical movement of the axlebox is only about 2 inches and is usually controlled by laminated leaf springs, either underslung or on top of the axlebox. The springs have about ten leaves and can vary in length up to a maximum of about 3 feet, depending on the size of the locomotive. Coil springs are common on bogie and pony trucks, and are occasionally utilised on the driving wheels of small tank locomotives.

A typical axlebox consists of two sections, split horizontally across the centre. The bottom half of the box is known as the underkeep and is usually made of brass or cast iron; it forms an oil bath for the lubricant. The top part of the axle box has a white metal liner cast on to a bronze or brass crown, forming a bearing surface for the axle to rotate and carry the weight of the locomotive. The liner continues to the outer face of the axlebox against the wheel centre in order to control the lateral movement, or end float, of the wheelset.

Wheel size

The size of the driving wheels gives a clue to the power of the locomotive and the purpose for which it was designed. If the

Top: The laminated spring fitted to the Cartazzi Bissel truck of an A4 class 'Pacific' locomotive. The photograph clearly shows the horn guides on either side of the axlebox and the horn stay beneath. *D. Dyson*

Bottom: Coil springs fitted to the Bissel truck of No 71000 Duke of Gloucester. The axlebox is of the roller-bearing variety. *D. Dyson*

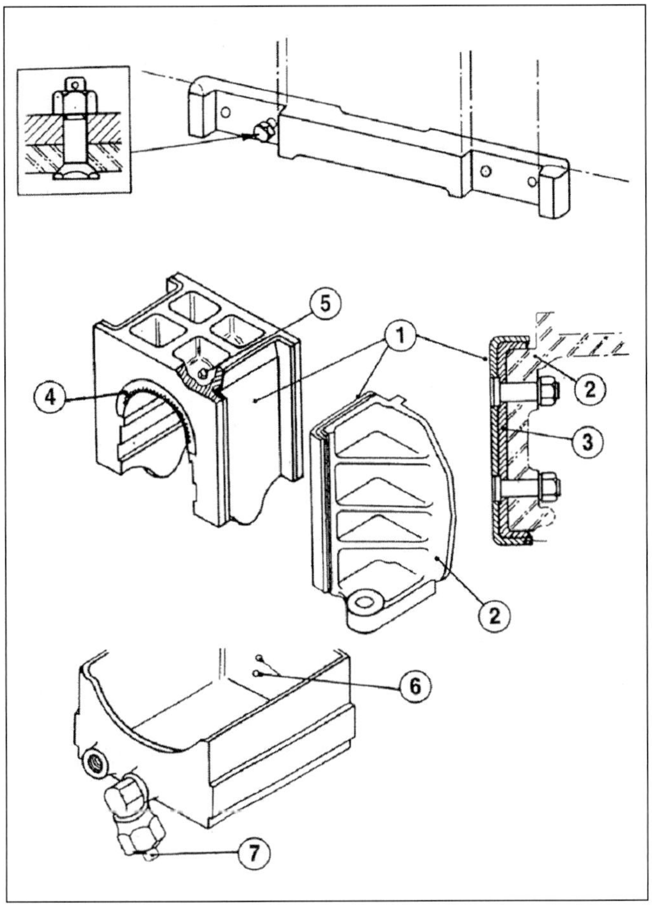

Left: A typical axlebox, showing (top) the horn stay, (centre) the axlebox and horn guide, and (bottom) the axlebox keep. In the axlebox can be seen the brass cap with the white metal liner and the slides for the horn guides. In the axlebox keep is the connection from the mechanical lubricator, while at the opposite end are oil holes to the end surf ace of the keep that controls the end float: 1 manganese steel liners; 2 horn guide; 3 spacer; 4 brass cap with white metal liner as bearing surface; 5 bolt ; 6 oil holes to felt; 7 feed from mechanical lubricator

Below: A frame stretcher between the horn guides: 1 stretchers running across the frames; 2 pins; 3 horn guides

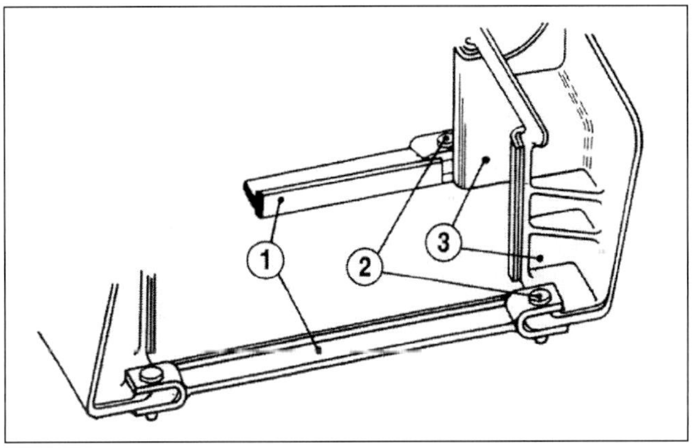

driving wheels are of a small diameter the engine is intended to pull heavy trains; if they are large it will have been built for high-speed running over long distances.

To illustrate this point I will use two entirely different locomotives designed and built for completely different purposes. The first is the two-cylinder 'Single' designed by Patrick Stirling and built in 1870 by the Great Northern Railway. This locomotive had a single pair of 8-foot-diameter driving wheels and was intended for high-speed non-stop running with a light load. To start a train with only one pair of driving wheels of this diameter on a greasy rail was difficult due to the low adhesion factor; there was even a problem with slipping at high speed. However, the large driving wheels meant that the demand for steam was not great, given the limitations of valve gear design at the time.

The second example is the unique four-cylinder 'Big Bertha' 0-10-0 No 2290 (BR No 58100) built by the Midland Railway in 1919. This famous ten-wheeled locomotive was used for many years for assisting trains on the famous 2-mile Lickey Incline between Bromsgrove and Blackwell, until it was withdrawn from service in 1956. The gradient of this incline is 1 in 37, and the ten driving wheels gave the locomotive sufficient adhesion to push heavy trains to the top.

Additional wheels and wheel arrangements

In the early years, small locomotives were sufficient for the traffic of the day, but it soon became evident that larger engines would be needed to handle the ever-increasing weight of trains and the longer distances being covered. As locomotives grew in power and size so did their weight, which caused damage to the track. It was soon realised by the railway civil engineers responsible for bridges and viaducts along the route that a limit for the weight carried by the wheels of locomotives would have to be enforced.

It was not possible simply to increase the number of axles and driving wheels to displace the weight because the locomotive had also to negotiate the various curves in the track. If the locomotive's wheelbase was increased to more than, say, four axles on a rigid frame, there was a tendency for it to straighten out the rails or for the engine to become derailed. Some eight-coupled locomotives such as the Gresley P2 class 2-8-2 for the LNER allegedly caused damage to the track, and were later rebuilt as 4-6-2s.

Thus the fitting of extra wheels in addition to the driving wheels helped to steer the locomotive, as well as carrying some of the weight and enabling the locomotive to travel at a higher speed and for greater distances. The combination of driving and other wheels led to what is known as the 'wheel arrangement' of the locomotive. The method of notation was laid down by an American railway engineer named Frederick Whyte, and takes the form of three numbers separated by a dash; a 'Pacific' locomotive, for example, is described as a '4-6-2'.

Looking at the side of the 'Pacific', it has at the front four small wheels (on two axles) fitted to a bogie that pivots; these wheels carry some of the weight and steer the front end of the

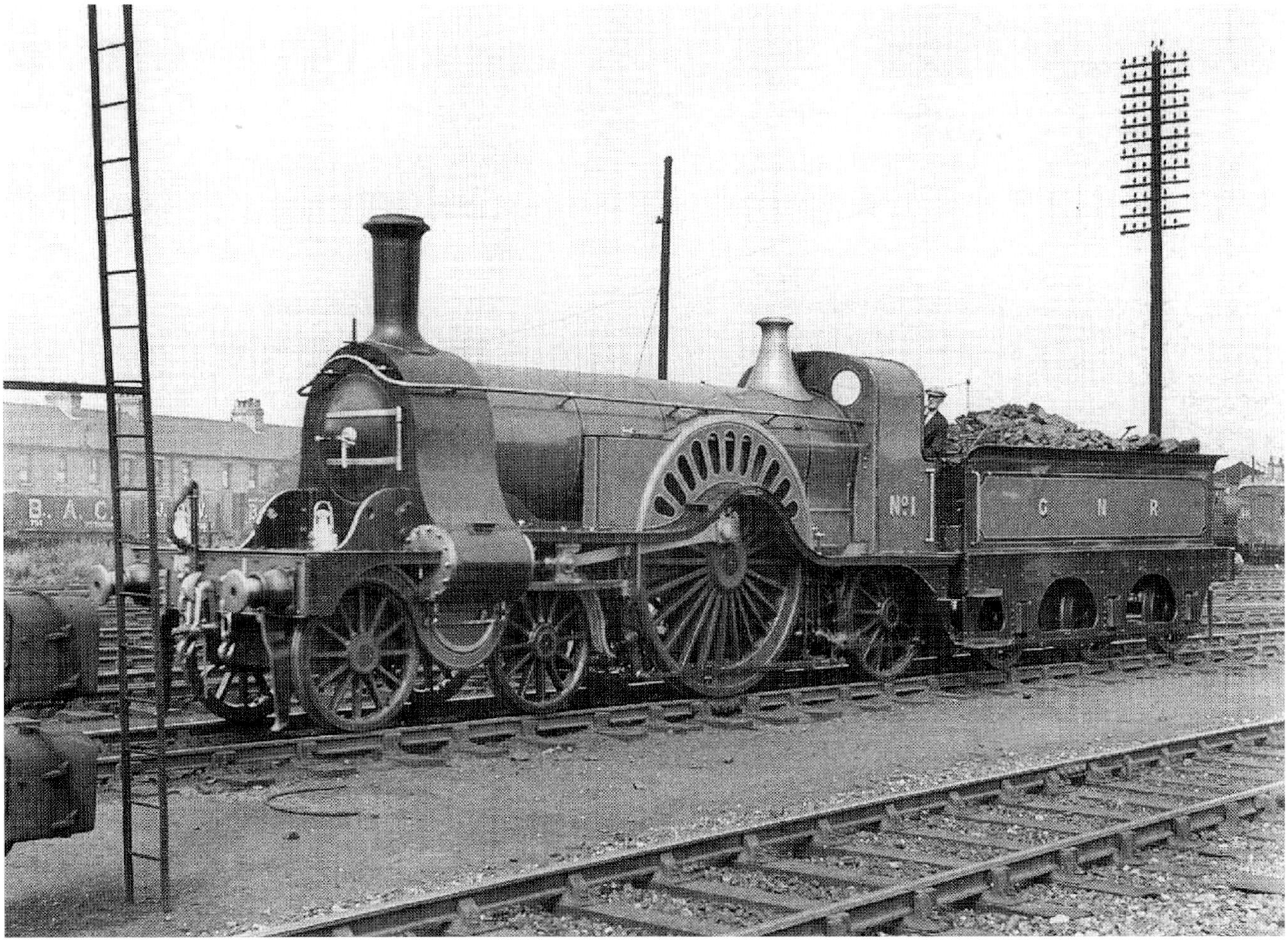

Above: Great Northern Railway Stirling 'Single' No 1, with its single pair of 8-foot driving wheels. (National Railway Museum, York)

Below: Midland Railway 0-10-0 'Big Bertha', its ten coupled driving wheels giving it maximum adhesion for assisting trains up the Lickey Incline. (National Railway Museum, York)

A Class 9F 2-10-0 locomotive - two wheels on the leading truck, ten driven wheels, and no wheels beneath the cab. *D. Dyson*

locomotive. Next are the six larger coupled driving wheels (on three axles), which deliver the power to the track by means of adhesion and drive the locomotive; they also carry much of the weight. Fitted just behind the driving wheels are two more smaller wheels on a single axle that also carry some of the weight and support the rear of the locomotive. These wheels can either be fitted to a pivoting bogie frame attached to the main frame, or mounted on the rigid frame of the locomotive in a type of horizontal sliding axlebox that allows the wheels to articulate. The latter is called a Cartazzi truck, and can be seen, for example, on LNER A3 class 'Pacifies'. The Cartazzi truck is similar in design to the radial axle box fitted to tank locomotives, only in reverse, with the frames outside the wheelset.

In France and Germany, the axles rather than the wheels are counted, so the wheel arrangement for a 'Pacific' would be '2-3-1', while in Germany the same locomotive would be a '2-C-l', the number of powered axles being denoted by the corresponding letter of the alphabet. The wheels on the tender do not form part of the wheel arrangement notation. However, the tender is fitted with brakes so the weight of the coal and water can be used to supply additional braking effect.

Tank locomotives

Tank locomotives have no tender to carry the coal and water, and their wheel arrangement notation 1s followed by the letter 'T'.

The design and shape of the water tanks varies, but the most common is the side tank, fitted to each side of the boiler and extending from the frames to about boiler height. On some side tank locomotives extra water is also carried in a tank below the coal space. Another popular type is the saddle tank, where a semi-circular tank is carried over the top half of the boiler. Some former Great Western locomotives employ pannier tanks carried high up on either side of the boiler. Some early types used a well tank, which carried the supply of water in between the frames.

Some tank locomotives required extra coal and water to allow them to work over longer distances, and these employ a longer wheelbase to spread the weight. They are fitted with a pony truck at one or both ends, although some were built on rigid chassis with no pivoting bogies. To allow these rigid-chassis engines to negotiate the tight-radius curves that can be found in goods yards or stations, the extra wheelsets are fitted with what is known as a radial axlebox. This allows the wheelset to pivot in the rigid main frames of the locomotive, the movement controlled by horizontal radial horn guides.

Radial tank locomotives were designed by F. W. Webb on the LNWR, the first being his 2-4-2T tank engines. In the Webb arrangement the two axleboxes of the outer pairs of wheels are connected together in an arc, but without any radial arm as with the conventional pony truck. These radial axleboxes are fitted inside the main frames of the locomotive, either at both ends, as in the 2-4-2T, or at just one end, as at the rear of ex-Great Western 2-8-2 tank locomotives. These entered service in 1910 as 2-8-0Ts with a pony truck at the front, but some were converted to the larger wheelbase when the coal bunker was enlarged to allow the loco to work longer distances, and a radial truck was fitted to that end.

	Britain	France	Germany
	0-4-0	0-2-0	B
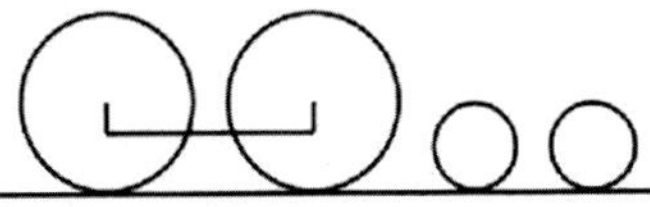	0-4-4	0-2-2	B-2
	2-4-0	1-2-0	1-B
	2-4-2	1-2-1	1-B-1
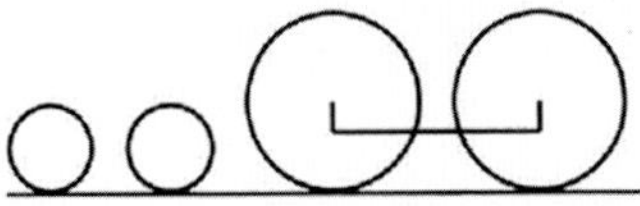	4-4-0	2-2-0	2-B
	4-4-2	2-2-1	2-B-1
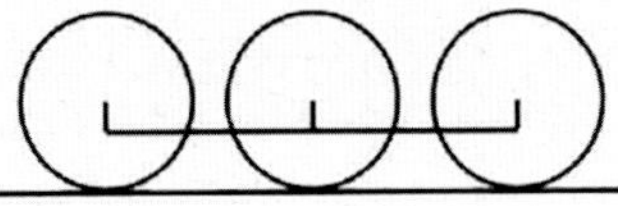	0-6-0	0-3-0	C
	0-6-2	0-3-1	C-1
	2-6-0	1-3-0	1-C
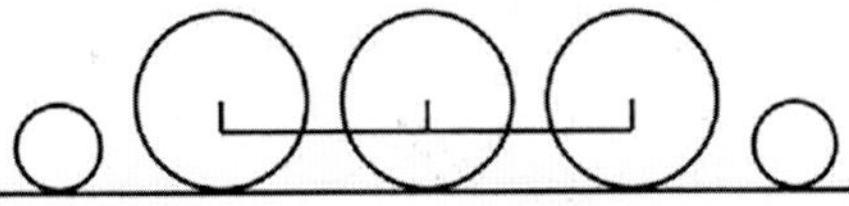	2-6-2	1-3-1	1-C-1
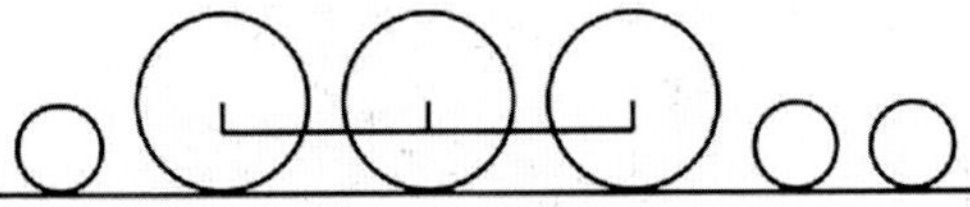	2-6-4	1-3-2	1-C-2
	4-6-0	2-3-0	2-C
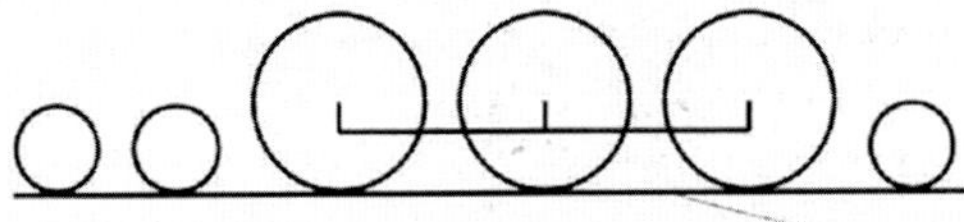	4-6-2	2-3-1	2-C-1
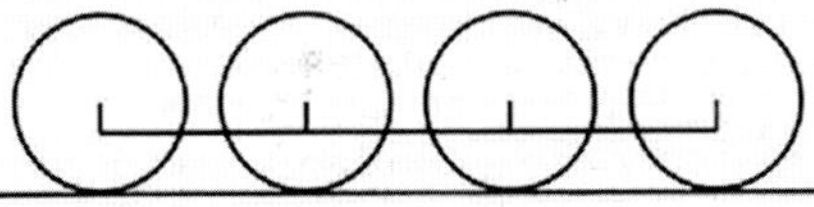	0-8-0	0-4-0	D
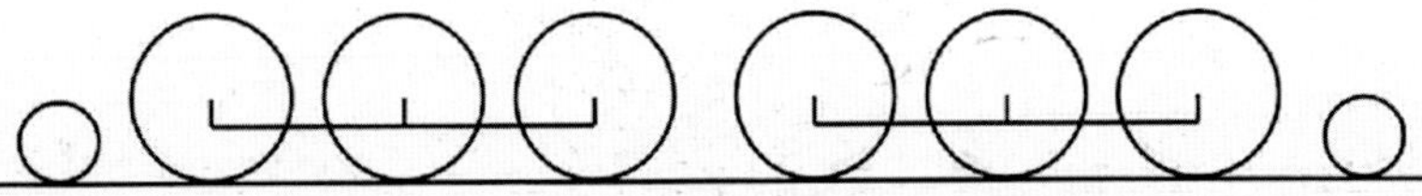	Garratt articulated locomotive 2-6-0+ 0-6-2	 1-3-0+ 0-3-1	 1-C+ C-1

Wheel arrangements - the Whyte notation and Continental equivalents.

Three examples of 0-6-0 tank locomotives designed for shunting and light traffic. Note the small diameter wheels, which would give plenty of power for short distances. The designs show side tanks (top), saddle tank (centre) and pannier tanks (bottom). *Author/Steve Davies collection (2)*

No 69999, the former LNER Garratt articulated locomotive with a power bogie at each end of the locomotive. *J. A. H. G. Coltas*

Articulated locomotives

An articulated locomotive such as the Garratt design is effectively two locomotives joined together with one boiler mounted on a common rigid chassis. At each end of the chassis or frame is a powered bogie able to pivot where the curvature of the track requires; each bogie has its own set of cylinders and motion working independently of the other from the common boiler.

The wheel arrangement notation for an articulated locomotive consists of six numbers representing both ends of the locomotive and separated by a 'plus' sign, for example '2-6-0 + 0-6-2'. No separate tender is used by this type of locomotive, which was designed to work heavy coal and iron ore trains on twisting routes. As with the tank engine, there was the additional benefit of most of the weight of the locomotive helping with adhesion.

A few types of articulated locomotive were built in England by the LMS and LNER, but none have survived into preservation. Several other designs were also tried, such as the French Meyer; similar to the Garratt, this employed one boiler with two pivoting power bogies that were linked together. Another example was the Mallet, which differed from the Garratt and Meyer in having the boiler and one set of cylinders carried on the rigid main frames while at the front was a pivoting bogie with another set of cylinders working from the common boiler. These locomotives were often 'compounds' (of which more below), with the high-pressure cylinders on the main frames and the low-pressure cylinders on the pivoting bogie at the front.

Another type is the Fairlie patent locomotive, which has achieved fame on narrow gauge railways. Like other articulated locomotives, it consists of a pivoting power bogie at each end of the frames, but under a double boiler with two fireboxes and a chimney at each end; the Fairlie is in effect two locomotives mounted back-to-back on one common chassis. The driver and fireman ride on the footplate in the centre cab, standing on opposite sides of the boiler. There are still some examples of these Fairlie patent 0-4-0 + 0-4-0 articulated locomotives in regular use on the Ffestiniog Railway in Wales, where the track is light with sharp curves and steep inclines.

Calculating power output

The power of a locomotive is usually described in terms of 'tractive effort' or 'drawbar horsepower'. The former is often wrongly interpreted as a measure of the effective power of a locomotive, but is really only a theoretical figure indicating the maximum pull that an engine can exert from stationary given perfect adhesion and tight glands and pistons, and has little to do with the actual work of which it is capable. For example, a J94 class 0-6-0 saddle tank has a comparable tractive effort to a 'Black Five' 4-6-0, but no one would suggest that it was as powerful!

The late Dr W. A. Tuplin, railway historian and engineer, describes tractive effort as 'the mean value (during any revolution of the driving wheel) of the backward push that the loco could exert on the rails when in full gear with maximum boiler pressure if the friction between coupled wheels and rails was great enough to transmit the pull and there was no friction anywhere else.' These conditions are never, of course, satisfied, and in any event tractive effort falls off immediately the locomotive begins to move, due to internal mechanical losses, etc.

Drawbar horsepower is a better measure, and is calculated from readings taken by a dynamometer car attached to a locomotive over a specially organised run with a known load and over a substantial period of time or distance.

Two firebars illustrating the length difference and their bow shape. At the ends and centre are the lugs that keep the bars separate to maintain a flow of air to the fire. *Brian Dobbs*

THE FIREBOX

The firebox, as the name suggests, contains the fire that heats the water, thus generating the steam that powers the locomotive. The size and shape vary from loco to loco. The grate area of the firebox of a 'Black Five', for example, is in the region of 28sq ft, while that of a small tank loco such as a pannier tank might be about 15sq ft or less.

Firebox components

Inside the firebox is the combustion chamber, forming the upper section above the firebed. At the base can be found the firegrate. Some grates slope towards the front, while some are flat; others are flat under the firebox door then slope away to the front under the brick arch. The grate is made up from cast iron bars placed in rows running from back to front. The bars may vary in length from about 2 to 4 feet, and they are carried on supports called 'feathers' that run across the base of the firebox. A small firebox may have only one row of bars, while a large one may have two, three or four rows.

The firebars are bow-shaped, about 1-inch-thick by about 4 inches deep at the centre, and have lugs cast into both ends and the middle. The lugs keep them apart to maintain an air gap of about three-quarters of an inch between the bars; if they were to close together with the vibration of the locomotive it would lead to starvation of air to the fire.

Some more modern locomotives are fitted with a rocking grate, so called because sections of the firebars can be opened from the footplate by levers to empty unwanted clinker into the ashpan for disposal at the depot, thereby giving the locomotive a faster turn-round between duties.

Under the firegrate is the **ashpan.** Its purpose is not only to collect the ash and clinker discharged from the fire, but it is also fitted with doors, known as 'dampers', which control the amount of air admitted to the firegrate, an efficient method of controlling the heat given off by the fire. In a small locomotive there may be just one damper door at the front of the ashpan, but most have front and rear doors fitted, while some have intermediate doors at the centre, depending on the grate size of the locomotive.

To aid the combustion of the fire a **brick arch** is fitted inside the firebox at the front, extending backwards at a slope just above the level of the fire; as its name suggests, it is made up from 4-inch refractory bricks to retain the heat of the fire and to prevent damage being caused to the tubeplate and the tubes should they come into direct contact with the

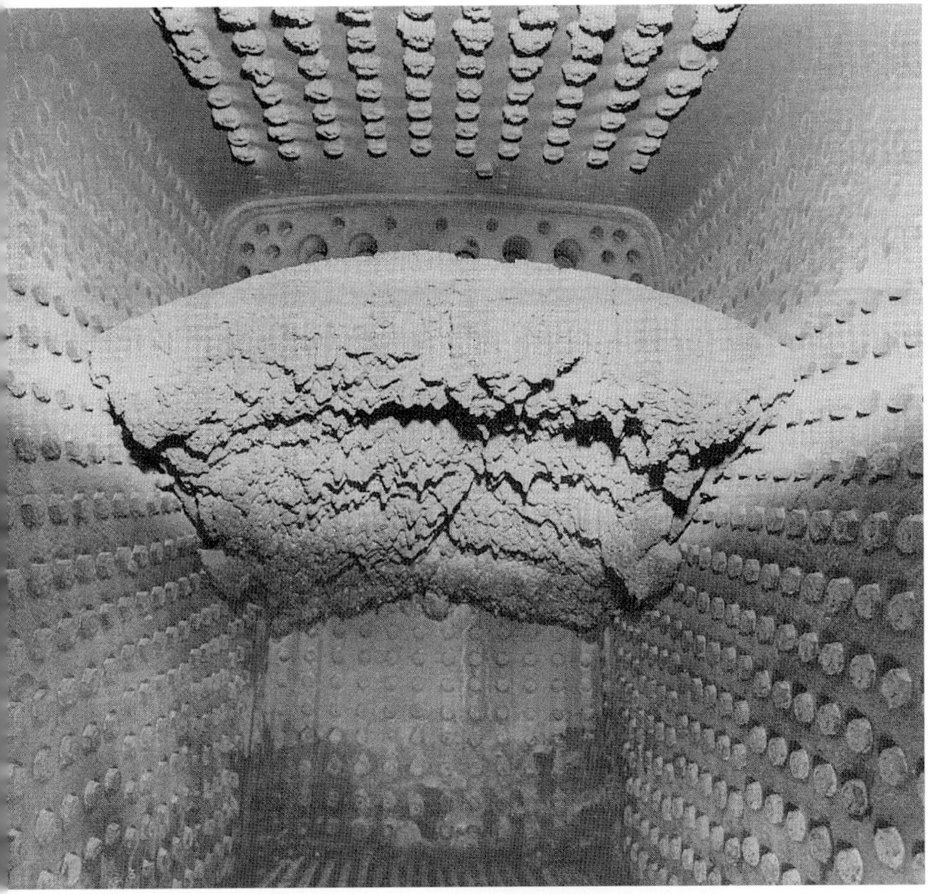

Left: The inside of a typical firebox, looking from the fire door. In the centre of the picture can be seen the brick arch, which has clearly seen considerable use. Above the brick arch the tubeplate is visible, together with the ends of the smoke tubes and the large flue tubes. Following the shape of the tubeplate, the lap joint between the firebox front, top and sides can be seen protruding into the firebox. On both sides are the nuts on the ends of the stays that protect the ends from damage from the fire; it will be noted that the ends of the stays above the line of the brick arch in the combustion chamber are just riveted over, as no damage will be caused at that height above the fire. At the bottom of the firebox are the rows of firebars that make up the grate. *Brian Dobbs*

Below: The arrangement of boiler plates between the outer firebox, on the right of the picture, and the top and bottom sections of the boiler; left. The shaped plate in between is known as the throatplate, which on this particular boiler is sloping. The domed-topped rivets securing the plates together are easily distinguished from the flat heads of the firebox stays. *R. G. Fox*

flames from the fire. Fitted inside the firebox door opposite the brick arch is a steel plate called a **baffle plate,** which acts in conjunction with the brick arch to aid the combustion of the fire. It also prevents cold air from being drawn on to the tubeplate, and is a safety feature helping to deflect the flame from the firehole door in the event of a 'blowback'.

Inner and outer fireboxes

The firebox itself consists of two boxes, inner and outer, separated at the bottom by a roughly 3-inch square piece of metal known as the foundation ring. Usually fastened to the foundation ring are the supports for the firebars. Built on to the inner edge of the foundation ring is a sheet of ¾ or ¾-inch steel or copper plate that forms the firebox inner wrapper plate. This plate is usually made up from one piece of metal and forms both the sides and roof of the firebox, although the roof is known as the crown sheet. At the front of the firebox is the tubeplate, which is secured by rivets to the wrapper plate. Opposite the tubeplate is the firebox door plate, also secured by rivets. The type of joint where the tubeplate and the firebox door plate meet the wrapper plate is called a lap joint, because the wrapper plate overlaps the outer edge of the tube and door plate by about 2 inches.

In early designs the inner firebox had a round top, leaving a relatively small combustion chamber. The Great Western Railway's G. J. Churchward altered the firebox to a square shape with sides sloping outward, giving a larger combustion chamber and therefore making more use of the heat given off by the fire. Making the outer firebox square to correspond with the inner also led to more water being carried round the firebox area, where most heat is available.

The inner and outer fireboxes are secured together by stays and nuts; due to the stress in the firebox, the stays are placed about 4 inches apart. On the firebox outer they are riveted over, while at the inner end they are threaded and fitted with nuts, which helps to prevent the ends being burned away by the fire. The stays fitted to the firebox crown sheet are known as crown or roof stays.

The firebox outer wrapper plate is again formed from one sheet of steel and is secured to the outer edge of the foundation ring by rivets. On the outer wrapper plate can be found the expansion brackets that secure the rear of the boiler in the frames. It is not possible to fix the boiler and firebox to the main frames of the locomotive at both ends with bolts because the boiler will expand with the heat when it is in steam. At the rear of the foundation ring can be found the firebox backhead or the outer firebox door plate. Fitted to the front of the wrapper plate is the throatplate, which joins the firebox and boiler barrel, also with lap joints.

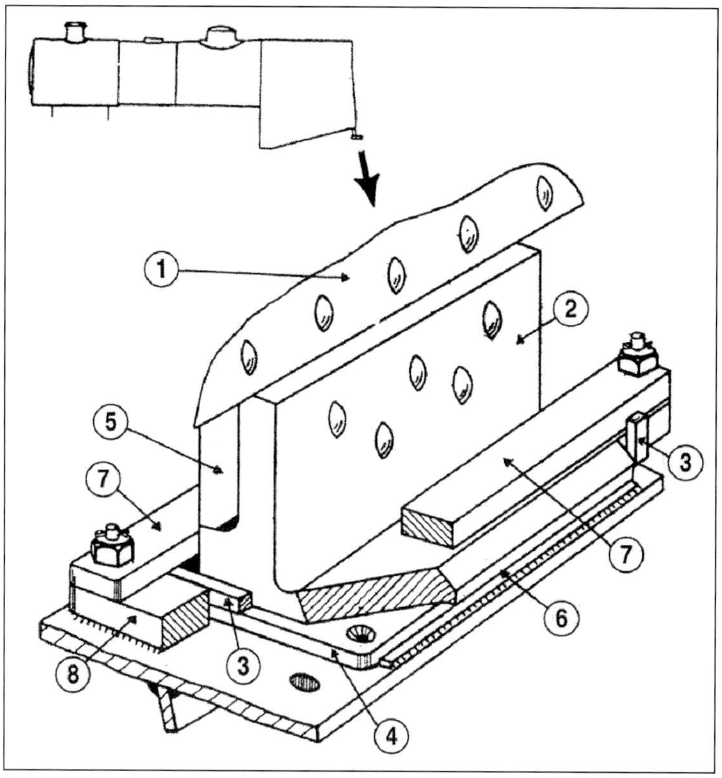

Details of the fixing plates of a Standard Class 4 locomotive that allow the boiler to move with expansion when the locomotive is in steam: 1 firebox backhead; 2 fixed boiler support foot; 3 side packing; 4 bronze slipper plate; 5 foundation ring; 6 frame cross-member below footplate assembly; 7 safety strap; 8 fixed side control

Fusible plugs

At no time when the locomotive is in steam must the firebox crown sheet or the tubes become uncovered by the water in the boiler, as this would cause serious damage. As a safety device to prevent this from happening, the boiler is fitted with fusible plugs, sometimes called lead plugs, screwed into the crown sheet of the firebox; there may be just one or, in the case of a large firebox, two or as many as six.

The plugs consist of a brass body about 1¼-inches square with a ½-inch or even a ¾-inch hole drilled right through, with a slight lip or bevel in the plug; the end of the plug is square so as to be easily distinguished from the nuts of the crown stays. The hole in the plug is sealed with a lead insert. Early fusible plugs had a thin brass cover over the lead insert and some were copper-coated, but these features were discontinued in favour of a dome-shaped top to protect the water side.

If the water level falls so low that the crown sheet of the firebox is no longer covered, the fire melts the lead insert and the steam and water left in the boiler escape into the firebox, thereby warning the crew of the defect. In a small firebox with a low fire the ingress of steam and water could possibly extinguish it, but in the firebox of a larger locomotive that is unlikely to happen. If possible both injectors should be placed on to feed water into the boiler, which will then pass through the blown plugs and into the firebox to help further with extinguishing the fire. It will also be necessary to remove the remaining fire from the firebox with the paddle or, in the case of a rocking grate, into the ashpan in order to prevent further damage being caused to the firebox. It perhaps goes without saying that such an eventuality would be a sign of *very* bad enginemanship!

Thermic syphon

Some boilers, particularly those of former Southern Railway 'Pacific' locomotives, had two large pipes or tubes running from the base of the tubeplate to the firebox crown sheet. These were known as thermic syphons and their purpose was to aid the circulation or convection currents of the boiler water.

Washing out the boiler

When a boiler is steamed, a muddy sediment is produced by the action of boiling water; this builds up and, by gravity, mainly congregates around the base of the firebox in the foundation ring area. Periodically this sediment must be removed; if it is left to build up, damage can be caused to the boiler and efficient steaming will be seriously impaired.

In order to facilitate the removal of this sediment, several 'mudhole doors' are fitted around the base of the outer firebox wrapper plates, while around the top of the firebox are similar plates, known as 'handhole doors'; both are usually oval in shape, about 4 inches by 6 inches, and can be identified by a single centre nut and a bridge-type clamp. They are kept steamtight and watertight by a gasket.

To assist in washing out parts of the boiler that are not otherwise accessible, and to allow sight of the stays, ere, square-headed nuts or plugs can be found about the boiler and firebox. These are fitted with a taper thread and can be removed in order to fill the boiler after a wash out or if the boiler is empty for some other reason.

The plugs and doors are also removed when the boiler receives its 14-monthly inspection by a boiler inspector. When a boiler is steamed, corrosion, caused by the salts in the water, takes place on the stays and tubes. Over a period of time the stays erode and become weak; they can also break due to the expansion and contraction of the boiler. When a boiler inspection is carried out all the doors and plugs are removed so that the inspector can carry out a visual check on the stays and tubes, if necessary with the help of lamps and long-handled mirrors. The thickness of the boiler plates will be sonic-tested for wasting by corrosion. If during the inspection a boiler fault is found, it will be condemned until repairs are effected. If any of the plugs or doors are passing steam when the boiler is in steam it must be reported to a competent fitter.

Above: Fusible plugs, showing their square shape and lead insert. *Author*

Safety valves

On top of the firebox can be found the safety valves. Most boilers have two, but the ex-LMS 'Pacific' locomotives have four because of the size of the boiler. Great Western locomotives had their safety valves mounted on top of the boiler barrel in a brass bonnet, which also contained the water feeds from the injectors, while steam was collected from the front upper corners on the firebox and passed to the regulator valve in the smoke box.

The safety valves fitted to early locomotives could be adjusted by the driver to give his locomotive more power; this was done either by screwing down the adjusting nut or simply hanging a weight on the arm to hold the valve on the seat. Needless to say, this led to some horrific boiler explosions and the practice was discontinued; safety valves fitted to a modern boiler cannot be tampered with by the driver.

Below: Cross-sections of a firebox fitted with thermic syphons running from the firebox tubeplate to the firebox crown sheet: 1 thermic syphons; 2 fire door - air is preheated as it passes through; 3 regulator valve; 4 superheater header; 5 main steam pipe; 6 multi-jet blastpipe

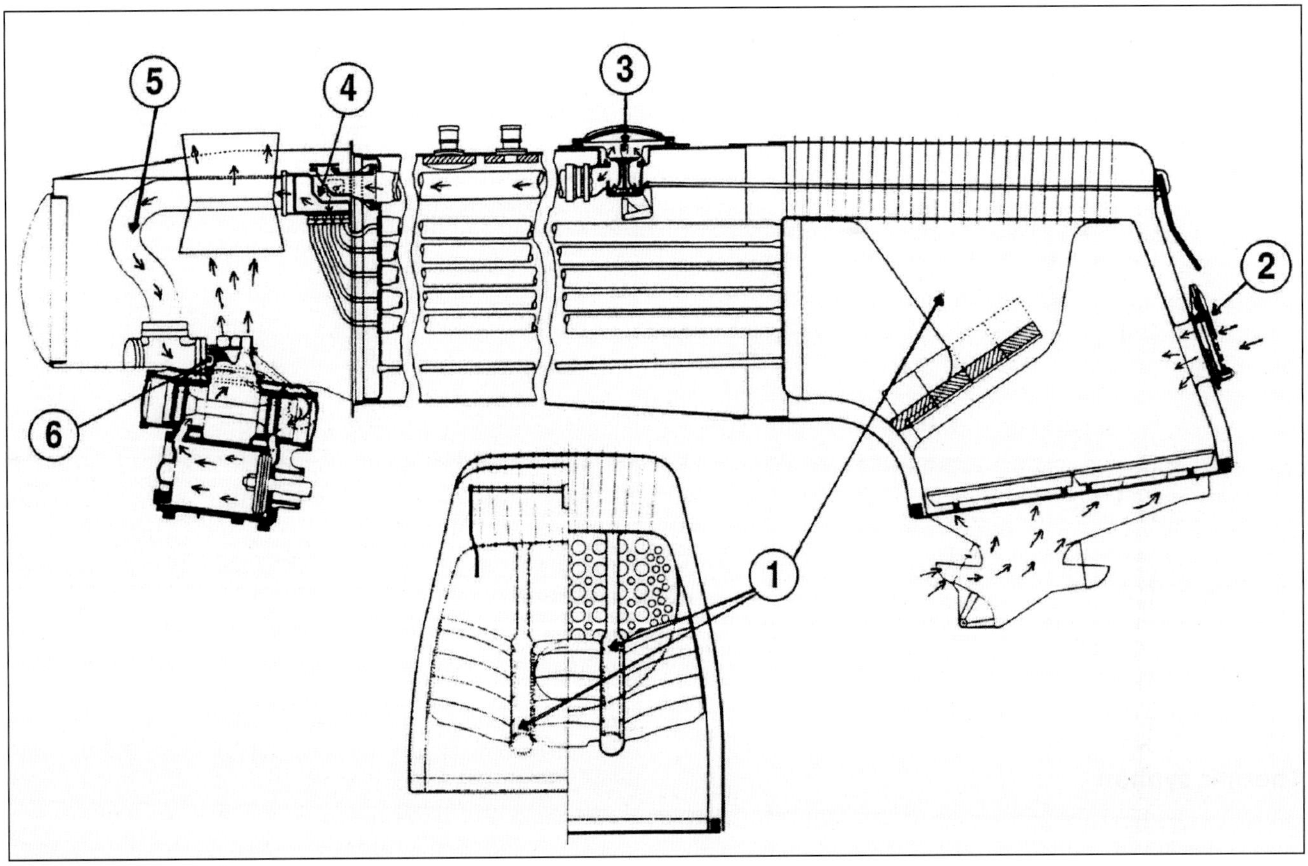

Above: A mudhole door at the base of a firebox - note the single nut and bridge-type clamp fitting. *Author*

Below: A pair of safety valves fitted to a BR Standard locomotive. *Brian Dobbs*

THE BOILER

Boiler design

The heart of a steam locomotive is the boiler, which varies in size and design from the early parallel type to the later taper boiler developed by the Great Western Railway in the early 20th century and adopted by other railways in various guises up to the end of steam.

The taper boiler was developed by G. J. Churchward, the GWR's Locomotive, Carriage & Wagon Superintendent from 1902 until 1921. He was the creator of the Great Western's locomotive standardisation policy, and the boilers developed by him were to become some of the most efficient producers of steam in the history of railways.

Another reason for the adoption of the taper boiler was to equalise axle weights. As the diameter of the boiler is smaller at the front, its weight is less, compensating for the weight of the cylinders. The taper boiler also improves the visibility of the line ahead for the driver and fireman, and when the locomotive is on a falling gradient the water is not able to surge to the front in such great quantities as with a parallel boiler, which might leave the water level low around the inner firebox.

The boiler barrel is made up from three, four or possibly five steel rings, depending on its length, design and shape. Some of these rings form a complete circle, while others, usually in the case of a taper boiler, are made up from semicircular plates riveted along their horizontal edges, then riveted to the rim of the next ring in a method similar to the lap joint. On top of the

boiler barrel is the dome, where the hottest and driest steam collects; the hotter and drier the steam, the better.

To retain the heat the boiler barrel and firebox are insulated with a jacket of glass fibre retained with wire mesh. In early types of locomotive and even well into the British Railways era the barrel was encased in asbestos; however, hazards to health caused by asbestosis have led to the use of glass fibre sheets, such as are used for loft insulation.

The insulation is then encased within metal sheets bolted to bands fastened to the boiler barrel and the outer firebox. On most locomotives the boiler backhead or faceplate is similarly lagged with steel sheet to retain the heat from the fire.

Boiler tubes

Running through the boiler barrel from the firebox tubeplate to the smokebox tubeplate are steel smoke tubes about 2 inches in diameter. These convey the heat from the fire through the water space in the boiler and take the burnt gases out to the atmosphere. A boiler such as that fitted to a 'Black Five' contains about 160 smoke tubes and 14 larger flue tubes. The surface area of these two sets of tubes is what

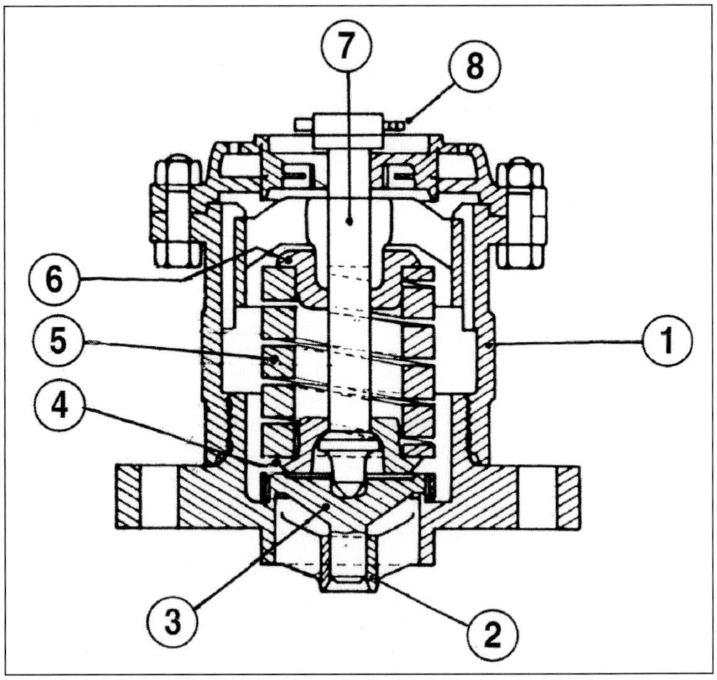

Above: A typical safety valve as fitted to a BR locomotive: 1 body; 2 valve guide; 3 valve; 4 bottom spring cap; 5 spring; 6 top spring cap; 7 stem; 8 pin

Below: A 'Crab' 2-6-0 with a parallel boiler (top) compared with a 'Black Five' 4-6-0 with a taper boiler, showing the difference in their shapes. *D. Dyson/D. Mathias*

Boiler smoke tubes waiting to be fitted into a boiler. Note that the ends are of a slightly smaller diameter to fit into the tubeplates. *Author*

is known as the 'heating surface', providing 1,463sq ft for the 'Black Five', although that figure may vary with the different types of boiler fitted to the class.

Superheating

Steam generated in a conventional boiler is known as 'saturated steam' and is not very efficient, so the Schmidt superheating company and others interested in increasing the efficiency of the steam locomotive developed the technique of 'superheating'.

Superheaters had first come into being in 1857, but were operated at relatively low temperatures; at that time there was no lubricant that could cope with more efficient higher temperatures, so development was suspended until continued by Dr Schmidt and others when lubrication improved.

The fitting of superheaters to some locomotives was not considered economical because of the additional cost incurred in construction. For example, they were not usually fitted to shunting locomotives or those that did not run for long distances because they did not achieve the required temperature of superheat on the short runs.

In the superheater the steam is re heated by being passed through a series of superheater elements contained inside the large tubes of the boiler, known as superheater flue tubes. These tubes are about 5 or 6 inches in diameter and run parallel to and above the small tubes from the firebox to the smokebox.

The steam is collected in the dome and passes through the regulator valve to the main internal steam pipe, or is collected in the dome by the steam collector pipe, from where it passes via the regulator valve to the superheater header in the smokebox. The superheater header, or steam collector, has two chambers side by side, the saturated side and the superheated side. The saturated side receives the steam from the dome, and it is then dried and further re-heated in the superheater elements running inside the flue tubes, which are connected back to the superheated side of the header.

As the steam passes through the elements in the flue tubes and is re heated, it increases in volume by about 30 per cent, giving it more elasticity and fluidity. The hotter and drier the steam becomes, the better, as this gives more efficient use of the fuel.

THE SMOKEBOX

At the front of the boiler barrel is the smokebox, the large chamber that forms the front end of the boiler. The smokebox is carried on a casting known as the smokebox saddle, which sometimes also forms the bottom of the smokebox. On some locomotives the saddle also forms part of the cylinder casting.

At the back of the smokebox is the tubeplate, carrying the ends of the tubes. In the case of a superheated locomotive the large flue tubes contain the superheater elements running to and from the superheater header or steam collector.

Running from each side of the superheater header are the two main internal steam pipes that run down to the cylinder steamchests. Where a locomotive uses saturated steam the

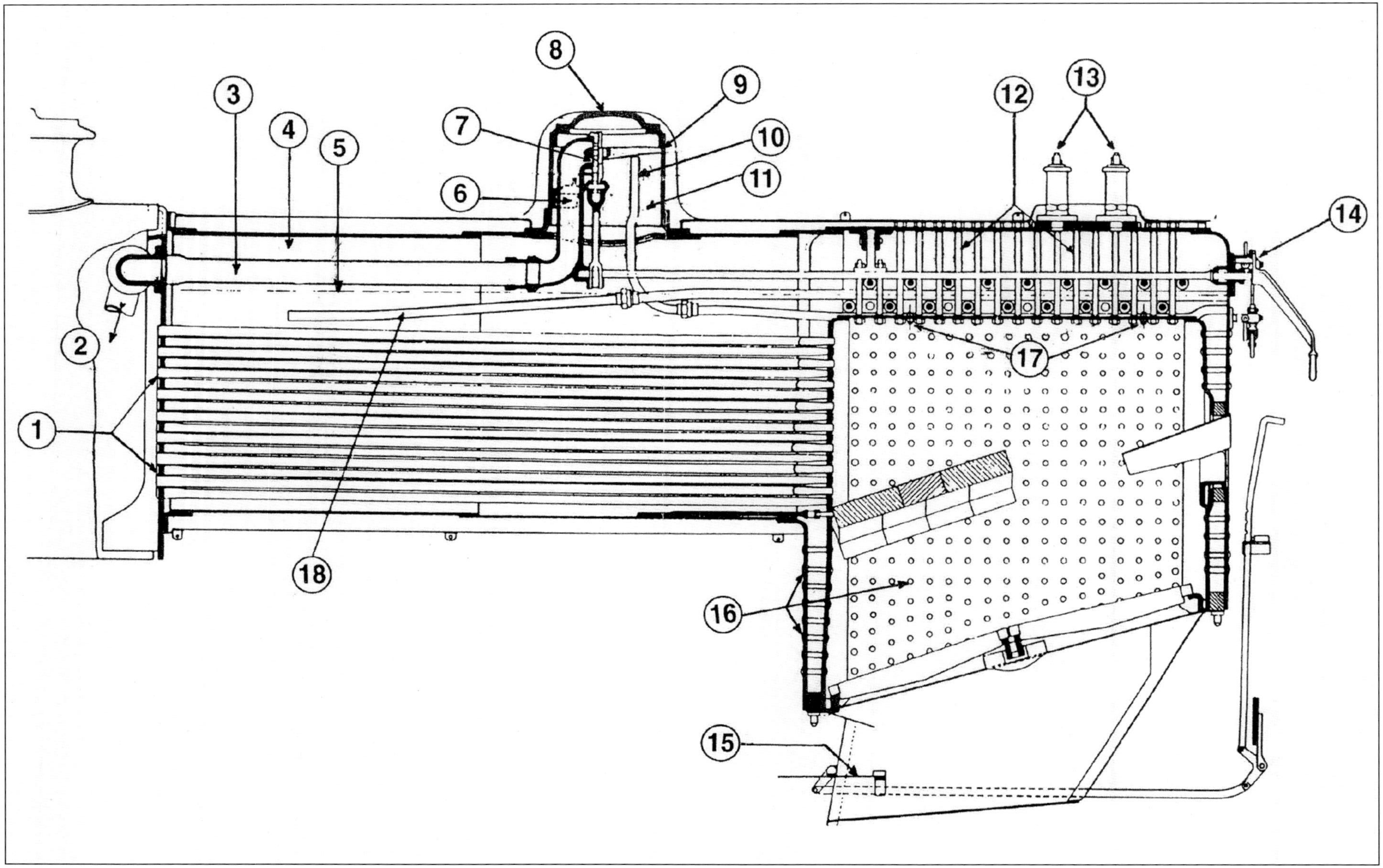

Sectional view of a firebox and parallel boiler, with no superheating. Note that the regulator valve is contained within the steam-collection dome: 1 smoke tubes; 2 steam pipe to cylinders; 3 main steam pipe; 4 steam space; 5 water level; 6 regulator head; 7 regulator main valve; 8 dome cover; 9 regulator pilot valve; 10 injector steam pipe; 11 dome; 12 steel roof stays; 13 pop safety valves; 14 regulator stuffing box and gland; 15 damper door; 16 copper side stays; 17 fusible plugs; 18 injector delivery pipe

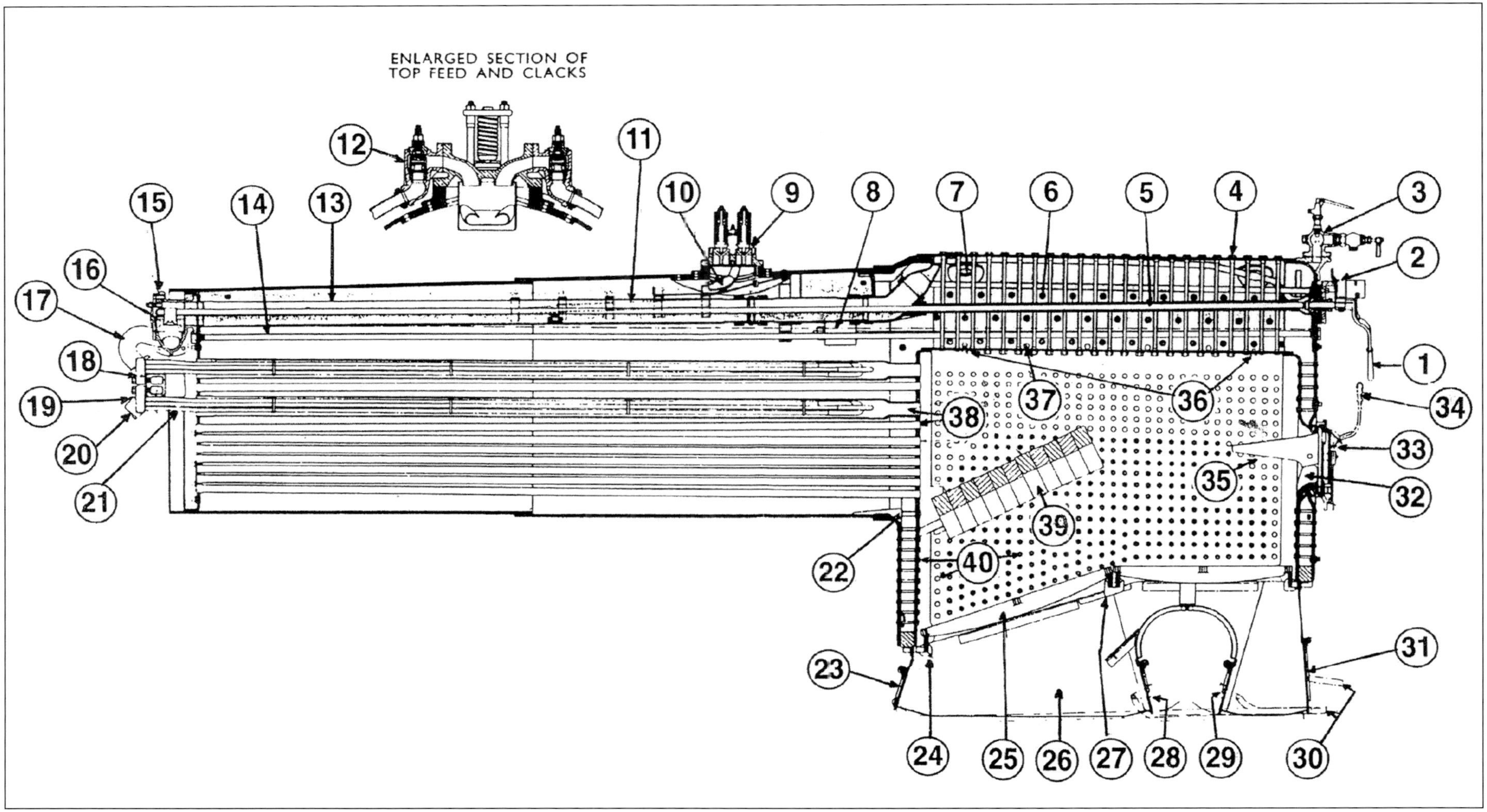

Sectional view of a firebox and taper boiler fitted with a superheater. Note that the regulator valve (16) is located in the smokebox: 1regulator handle; 2 regulator guide and stuffing box; 3 steam fountain; 4 crown stays; 5 regulator rod; 6 transverse stays; 7 washout door; 8 water level; 9 safety valves; 10 top feed water scoop; 11 top feed water trays; 12 clack box; 13 main internal steam pipe; 14 longitudinal stay; 15 regulator valve box; 16 regulator valve; 17 steam pipe, regulator to superheater; 18 main superheater header; 19 junction headers; 20 steam pipe to cylinders; 21 superheater elements; 22 palm stays; 23 front damper door; 24 firebar bearers; 25 firebars; 26 ashpan; 27 firebar bearer brackets; 28 front middle damper door; 29 back middle damper door; 30 damper gear operating rods; 31 back damper door; 32 firehole shield; 33 firehole doors; 34 firehole door handle; 35 baffle plate; 36 fusible plugs; 37 washout plugs; 38 large and small smoke tubes; 39 brick arch; 40 small water space stays

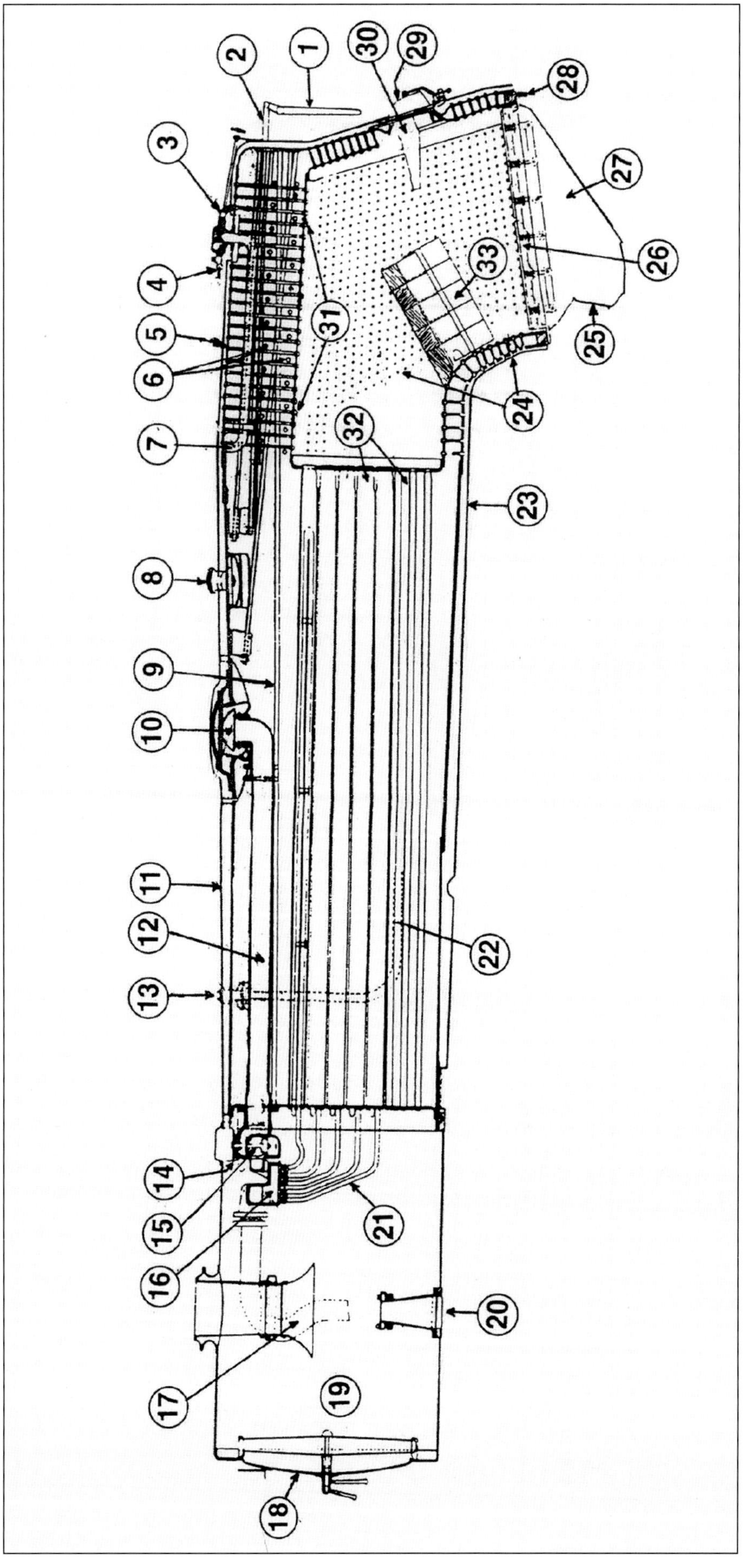

Sectional view of a modern firebox and superheated boiler of a British Railways Standard locomotive: 1 regulator handle; 2 regulator rod; 3 steam manifold; 4 whistle; 5 roof stays; 6 transverse stays; 7 steam manifold collecting pipe; 8 safety valves; 9 longitudinal stays; 10 steam collector; 11 boiler casing; 12 main steam pipe; 13 top feed connection and clacks; 14 regulator valve body; 15 multi-valve regulator; 16 superheater header; 17 steam pipe to cylinders; 18 smokebox door; 19 smokebox; 20 blastpipe; 21 superheater elements; 22 delivery pipe; 23 boiler casing; 24 water space stays; 25 front damper door; 26 firebars; 27 ashpan; 28 firebox steady bracket; 29 sliding firedoor; 30 baffle plate; 32 fusible plugs; 33 large and small smoke tubes

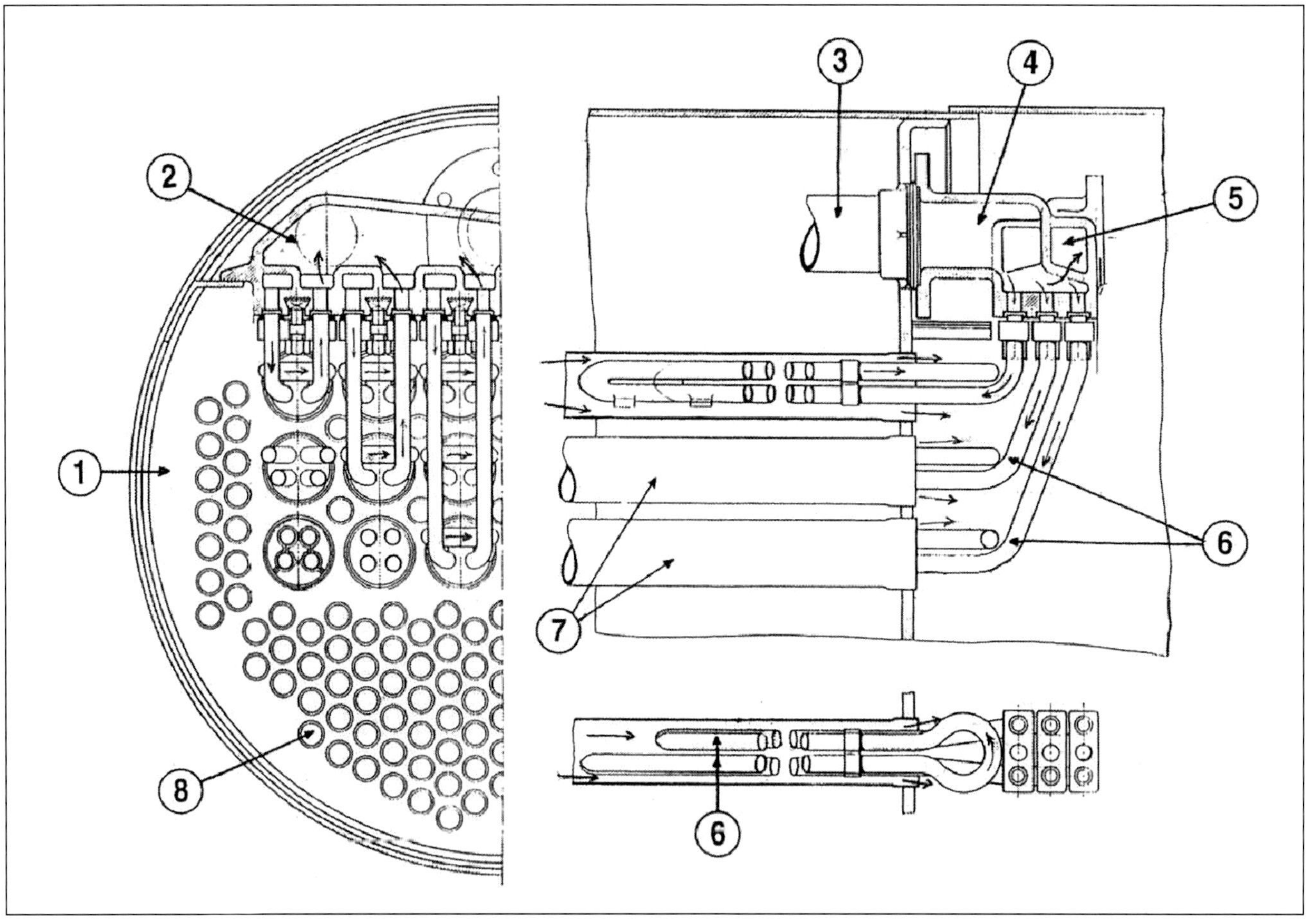

Arrangement of a smokebox superheater header and elements in the boiler flue tubes: 1 smokebox; 2 superheater; 3 main internal steam pipe; 4 superheater header (wet side); 5 superheater header (dry side); 6 superheater elements; 7 superheater flue tubes; 8 smoke tubes

main internal steam pipes run direct from the boiler via the regulator valve to the steamchests.

Blastpipe and chimney

At the base of the smokebox is the blastpipe, and directly above it and in line with it is the funnel-shaped petticoat fitted to the base of the chimney. The petticoat is specially shaped to work in conjunction with the blast of the escaping steam, drawing the combustion gases along the tubes from the firebox. The increased draught caused by the escaping steam makes the fire burn with intense heat and combusts the gases given off.

Fitted to the top of the blastpipe is the blower ring, which uses steam from the boiler via a valve on the footplate to simulate the exhaust blast and create a partial vacuum in the smokebox when the locomotive is stationary, to prevent a blow-back from the fire through the firebox door on to the footplate.

During the development of the steam locomotive several designs of blastpipe were tried, including the double blastpipe combined with a double chimney, intended to save coal by giving the locomotive better steaming qualities. One such is the 'Kylchap', developed by locomotive engineers Kylälä and Chapelon. This blastpipe was used by Andre Chapelon on his French 4-8-4 express locomotives, and employed in Britain by the LNER and the Southern Railway, as well as by British Railways on its Standard Class 8 express passenger locomotive No 71000 *Duke of Gloucester* Another attempt at fuel economy was the Giesl ejector, which, despite its name, has nothing to do with vacuum brakes. It was designed and developed by Dr Giesl-Gieslingen, but was not popular in Britain. A locomotive fitted with the Giesl ejector can be distinguished by its rectangular chimney.

Self-cleaning smokeboxes

Later locomotives have self-cleaning smokeboxes that are fitted with wire mesh screens so that the blast of the exhausting steam causes the clinker to blow about the smokebox. This violent action breaks up any lumps of clinker into ash, which is then exhausted by the blast up the chimney into the atmosphere. Such locomotives can be identified by a plate marked 'SC' on the smokebox door.

FILLING THE BOILER

Injectors

The need to maintain the water level in the boiler of a steam locomotive gave the early locomotive engineers some problems. Several methods were tried before the first successful injector was developed by the Giffard Company in 1858. Henri Giffard started his career on the Belgian State Railways and was a motive power depot foreman when he developed the injector.

It is important that there are two independent methods of filling the boiler, either two injectors or one injector and a pump. There are a number of different types of injector in use: some are operated entirely by live steam taken from the dome, while others are worked by both exhaust steam, taken from the base of the blastpipe in the smokebox, and live steam from the boiler. The latter type is known as an exhaust steam injector and employs a shuttle valve that operates automatically; when the regulator is closed the shuttle valve turns the injector back to live steam only. The reason for the use of an exhaust steam injector is economy; steam that would otherwise be wasted up the chimney is utilised to feed water into the boiler.

Both live steam and exhaust steam injectors work on the same principle, and require three types of energy to overcome the boiler pressure: velocity, momentum and pressure. A simple injector is made up from three specially shaped cones:

1 Steam cone, which gradually reduces in diameter and develops the required velocity.
2 Combining cone, where the steam is combined with the water, which gives the momentum.
3 Delivery cone, which develops the pressure in the water to overcome that in the boiler.

In all injectors, whether mounted vertically on the faceplate, or vertically or horizontally under the footplate steps, the cones are in line with each other.

The simple injector is operated by the following sequence of events. First the water handle is operated from the footplate to admit feed water from the tank or tender into the injector body until water can be seen running from the overflow; this expels any air from the body of the injector. When the steam spindle is opened, steam enters the injector via the steam cone. This cone gradually reduces in diameter to create the velocity required to operate the injector.

Details of the inside of the smokebox of 'Black Five' No 5407. At the bottom can be seen the blastpipe, and to the right the pipe from the blower. The pipe running to the casting just above the petticoat at the base of the chimney is the exhaust from the ejector. Just behind can be seen the superheater header with the two main internal steam pipes running down on either side to the steamchests; below the pipes are the superheater elements in the flue tubes, and beneath them the smoke tubes. *D. Dyson*

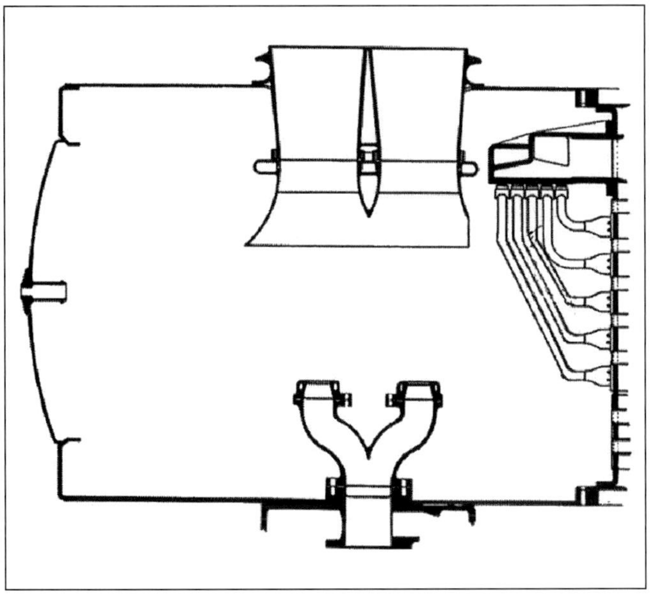

Right: The Giesl ejector fitted in the smokebox of BR Standard No 78022, as modified at the Keighley & Worth Valley Railway. Note the rectangular blastpipe leading to the base of the similarly shaped chimney. *John Sagar*

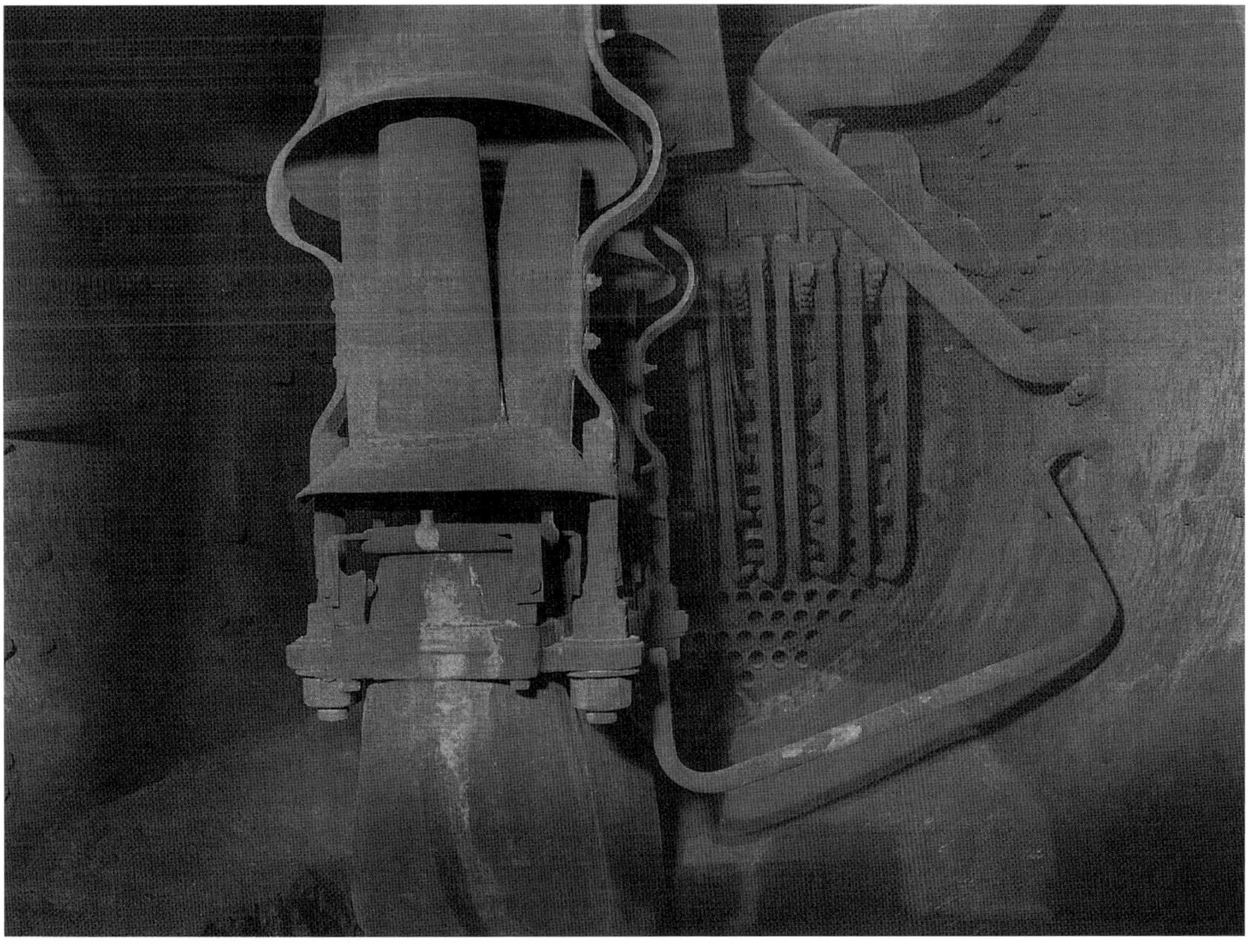

Inside the smokebox of No 71000 Duke of Gloucester, showing the Kylchap blastpipe; note the nozzles that take the blast up the double chimney. At the back of the smokebox can be seen the tubeplate and superheater elements. *Brian Dobbs*

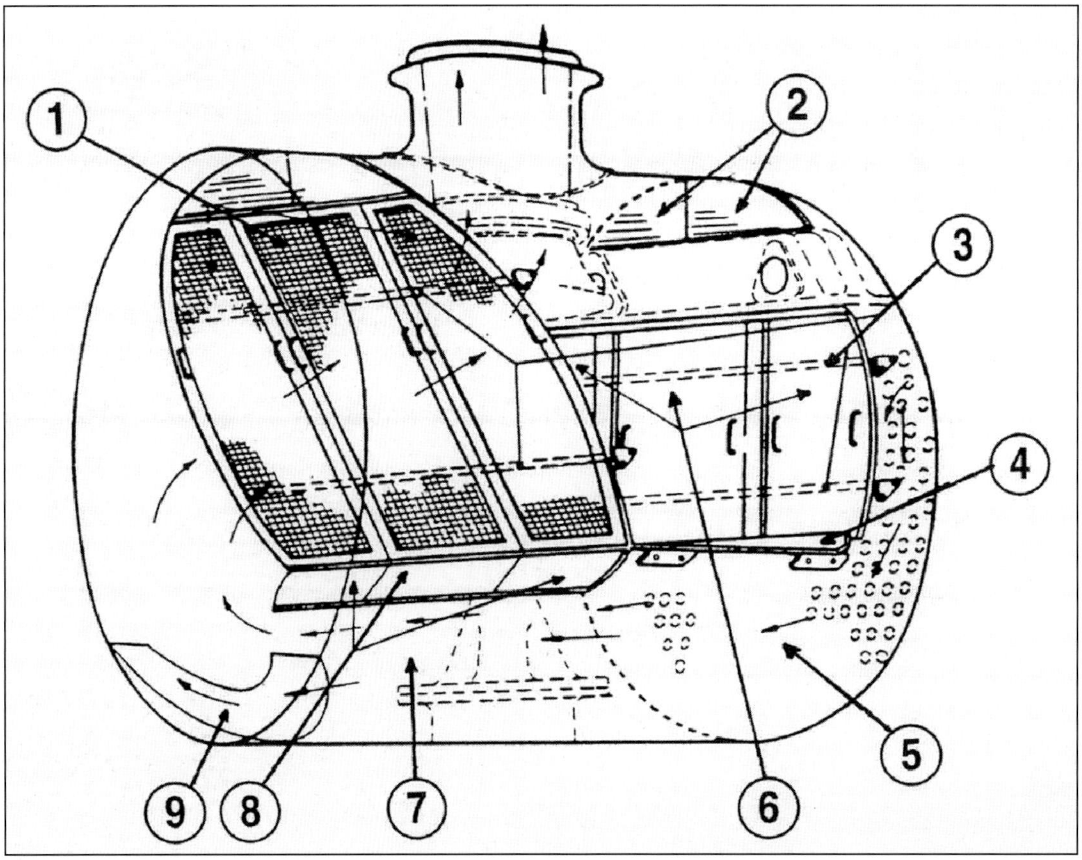

A typical self-cleaning smokebox. The wire mesh screen breaks up the clinker into small particles inside the smokebox so that it can be exhausted by the blast up the chimney: 1 mesh screen; 2 diaphragm plates; 3 supporting bars; 4 table plates; 5 tube plates; 6 diaphragm plates; 7 restriction caused by deflector plates for lifting ash; 8 deflector plates; 9 supporting bars

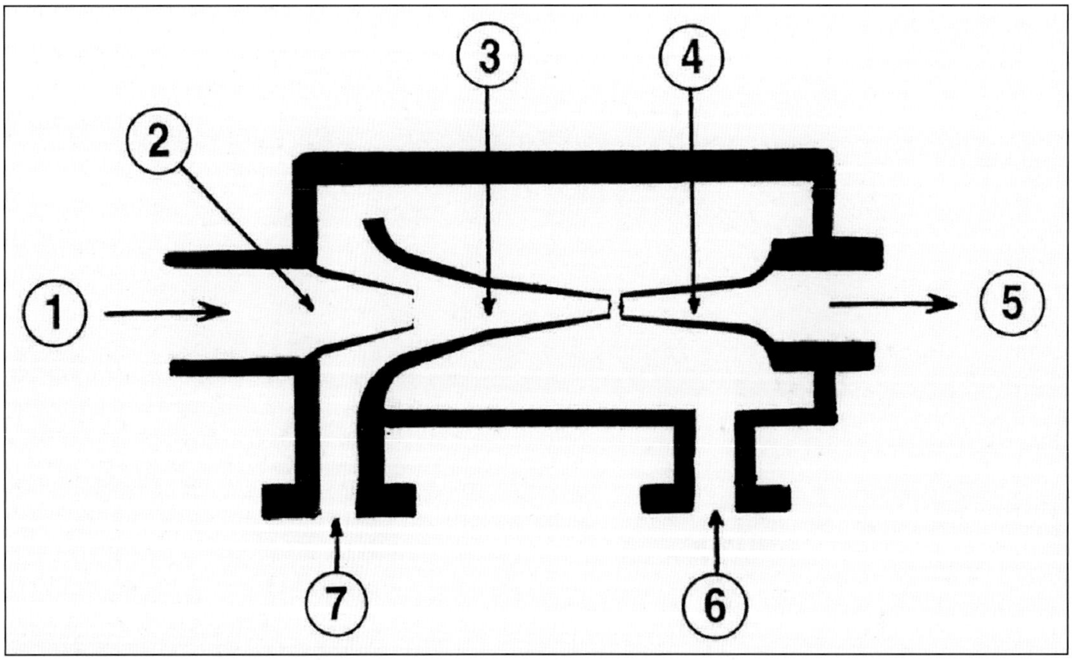

The simple injector. Steam enters through the steam cone (1), meets with the water in the combining cone (3), mixes with the water, condenses and becomes a solid pillar of water mixed with steam sufficient to jump the gap into the delivery cone (4). It then has enough force to overcome the boiler pressure: 1 steam flow; 2 steam cone; 3 combining cone; 4 delivery cone; 5 to top clacks; 6 overflow; 7 water supply

Both injectors mounted on the side of a British Railways Standard locomotive. It was normal to put one injector on each side of the locomotive, but as the designs improved both the injectors and their controls were moved to a convenient position on the fireman's side of the footplate. *Brian Dobbs*

The steam, having reached a speed of about 1,700 feet per second, is discharged into the combining cone, where it is completely surrounded with water and immediately condenses. Because of the velocity already created in the steam cone, the water must follow the steam to jump the gap and enter the delivery cone. This has an increasing diameter, creating pressure that is sufficient to lift the water up to the top clack and overcome the boiler pressure.

Most of the early devices suffered from interruptions caused by vibrations from rail crossings, etc., causing the injector to stop working; both the steam and water would then have to be shut off before the injector could be re-started, often with difficulty.

Automatic re-starting injectors were developed by, among others, the Manchester firm of Gresham & Craven and Davies & Metcalfe. These employ a special combining cone that overcomes the problem by being fitted with a hinged flap that opens to divert the pressure of the injector to the overflow if anything happens to interrupt the flow, then re-starts the injector. When the injector is working the hinged flap of the combining cone is kept closed by vacuum.

Another type of combining cone employs a moving cone that works on a similar principle to the flap. When the supply is interrupted, the moving cone separates to release the pressure into the overflow; as soon as the injector starts to work properly, the two parts of the cone are brought together by vacuum.

Top clacks

The water always enters the boiler through non-return valves called top clacks. These are positioned away from the firebox to avoid damage that might be caused by feeding cool water on to the hot plates. When the injectors are shut off the top clacks close automatically, although some can have

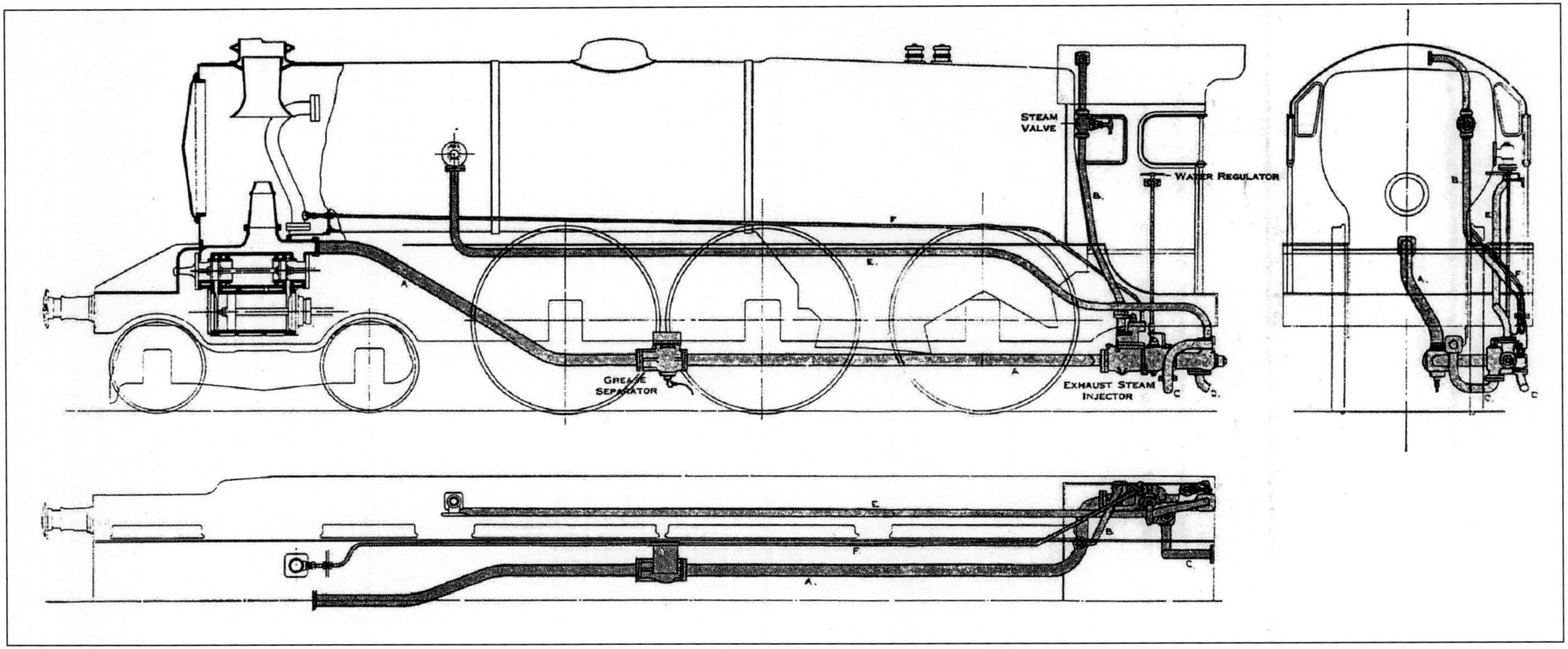

Above: Arrangement of the Class 'H' Davies & Metcalfe exhaust injector, with independent check and steam valves on the locomotive. (Davies & Metcalfe): A exhaust steam pipe; B live steam pipe; C feed water pipe; D overflow pipe; E delivery pipe; F connection to engine steam pipe

Right: The Davies & Metcalfe Class 'H' exhaust steam injector fitted below the footplate of a locomotive. Note the exhaust steam supply pipe at the front of the injector body. *Author*

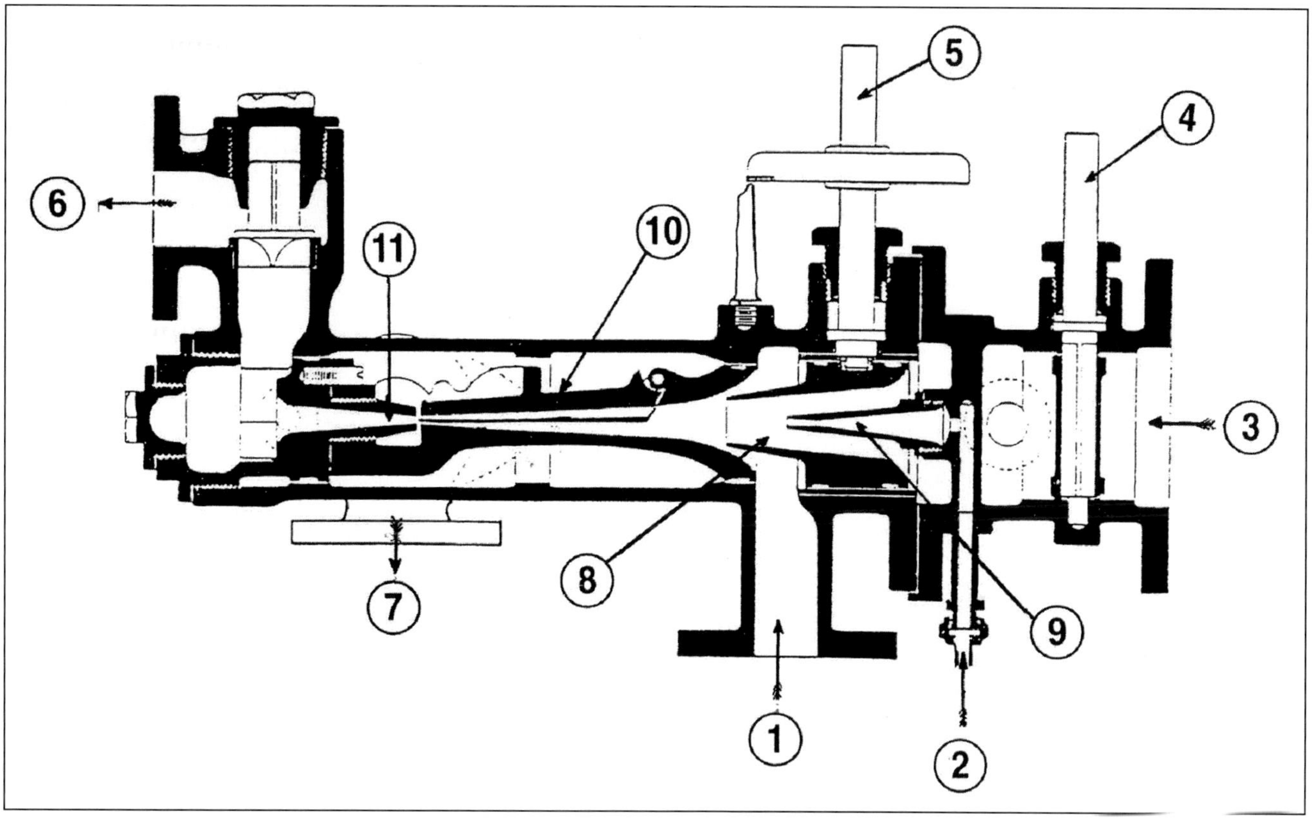

Above: The Davies & Metcalfe 'A' Type exhaust steam injector, showing the hinged flap combining cone. In the diagram the flap is in the closed (normal) position, and is kept closed by vacuum when the injector is working: 1 water inlet; 2 supplementary live steam inlet; 3 exhaust steam inlet; 4 exhaust steam valve; 5 water regulator; 6 delivery outlet; 7 overflow; 8 exhaust steam cone; 9 live steam cone; 10 combining cone with hinged flap; 11 delivery cone

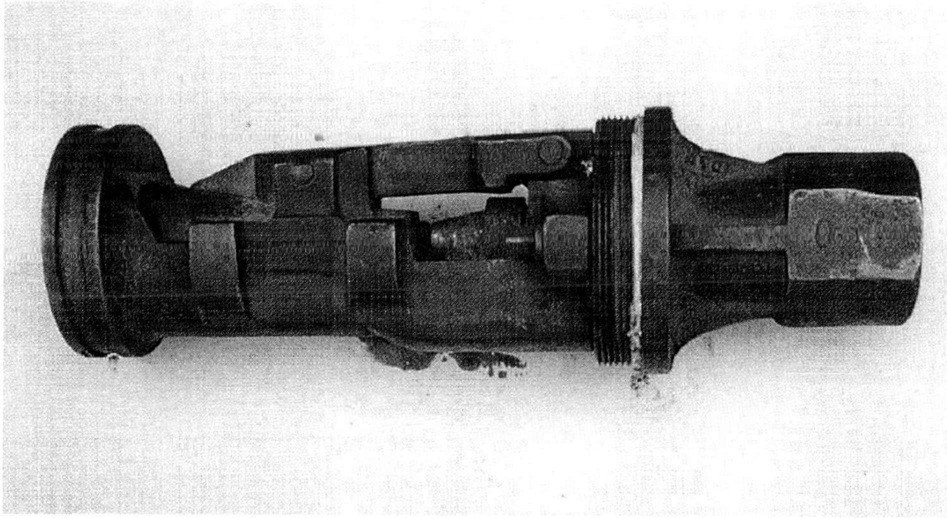

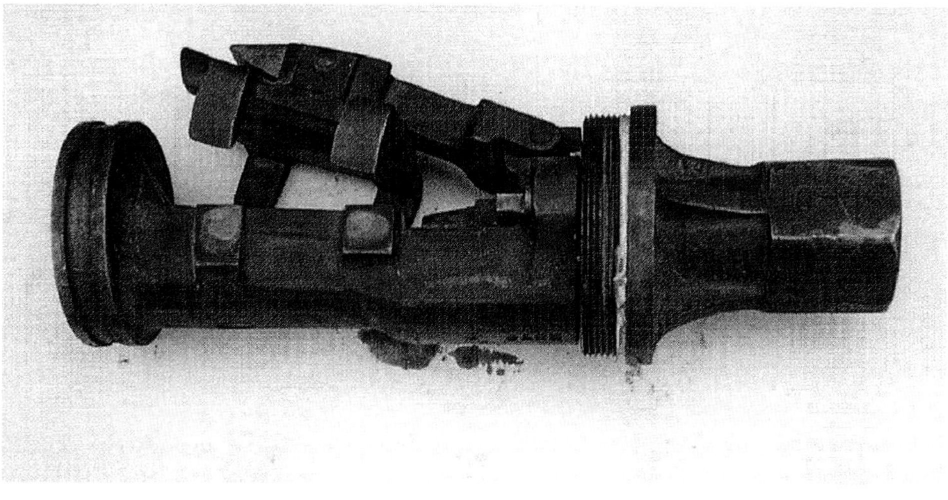

Right: The Davies & Metcalfe injector cone with hinged flap in the closed and open position. The flap is kept closed by vacuum when the injector is working, caused by water passing through the cone. It opens to relieve the pressure should the injector blow back. *Author*

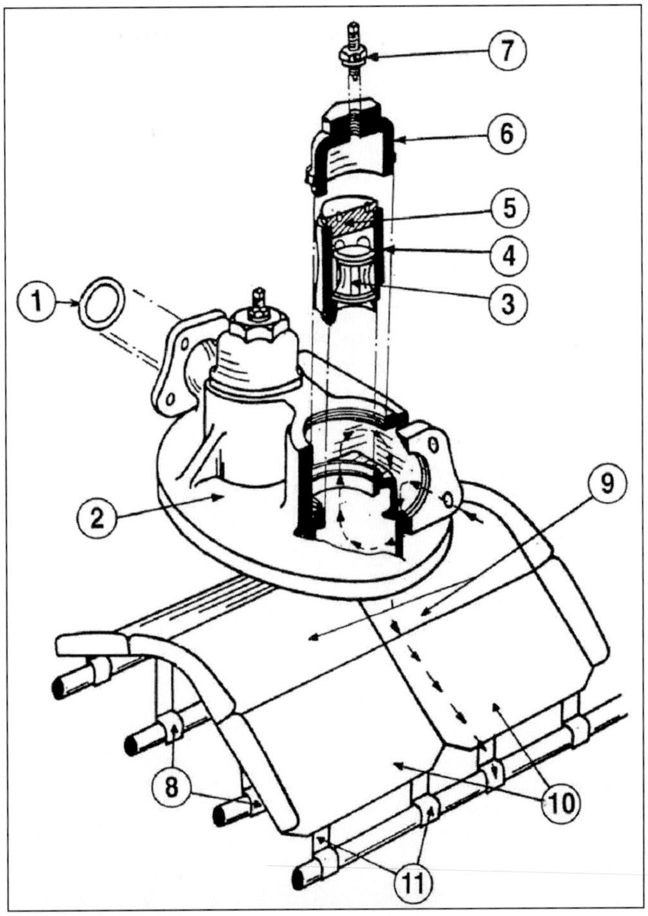

Above: Diagram of the top clack arrangement of a BR Standard locomotive. The water flows into the boiler in the direction of the arrows.

1 copper joint ring
2 clack body
3 clack
4 clack cage
5 clack cage cap
6 cap
7 stop screw nut and washer
8 carriers
9 crown plate
10 side plates or trays
11 supports

Above left: The Great Western's top feed arrangement is built into the safety valve bonnet. *Author*

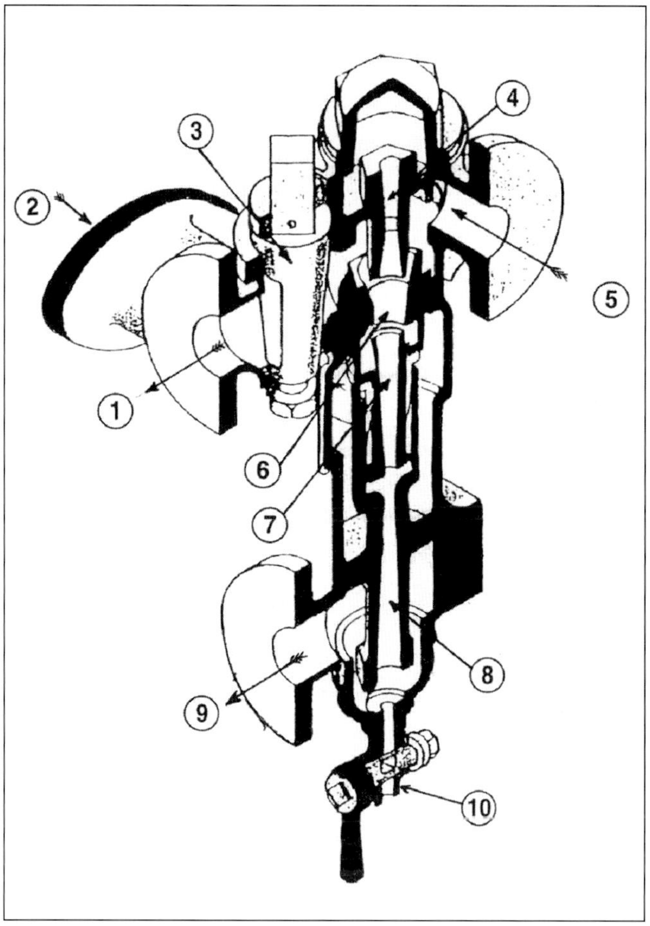

Above: A standard vertical injector showing the steam, water and delivery connections. The sliding cone in the centre of the injector performs the same purpose as the hinged flap: 1 water overflow; 2 water inlet; 3 watercock; 4 steam cone; 5 steam inlet; 6 fixed part of combining cone; 7 moving part of combining cone; 8 delivery cone; 9 outlet to top clack; 10 drain cock

Above right: A vertically mounted injector below the footplate of an ex-LMS locomotive. *Author*

Right: The gauge frame fitted to a BR Standard locomotive showing the gauge glass tube within its steel and glass protector, which was developed from the Great Western design. *Brian Dobbs*

an additional manual shut off on the footplate in case they blow back.

Gauge glasses

The level of the water in the boiler can be seen in the gauge glass fitted to the firebox backhead; water will always find its own level, so the glass repeats the boiler water level. There are two types of gauge glasses fitted to the boiler of a steam locomotive.

The first utilises a pillar of water in a glass tube known as a gauge glass; the glass is encased in a gauge frame should the glass shatter under pressure.

The second is the Klinger reflex water gauge, which can be found on locomotives of Continental origin and also on Bulleid 'Pacific' locomotives. The advantage of the Klinger type is that there is less danger of it blowing out. It uses a thick glass pl ate mounted in a metal body with prismic grooves on the water side. The grooves are arranged so that light strikes the glass in such a way as to show the steam space as bright and silvery, and the water space as black.

Continuous blowdown valves

The purpose of the continuous blow down valve, which may be fitted to the backhead or faceplate of the boiler, is to keep the boiler clear of the salts in the water that cause the locomotive to prime and cause scaling and corrosion that shorten the life of the boiler.

A small quantity of water, just over a gallon per minute, is taken from above the firebox crown and discharged into the ashpan, taking with it the waste boiler salts. The valve works continuously when the regulator is open and is actuated by a pipe from the steamchest. It also has a connection to the injector, which operates a ball valve when the miector is

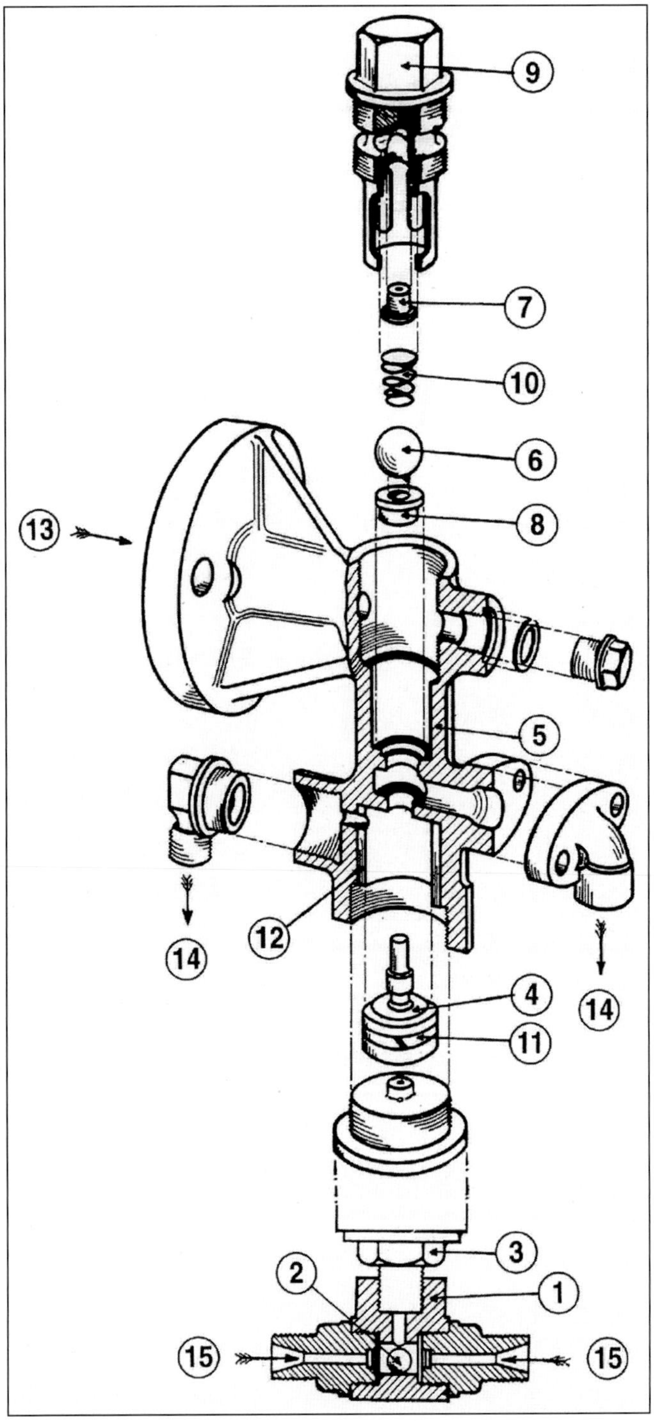

Above: An exploded view of a continuous blowdown valve showing the connections to the injector and the drains to the ashpan:
1 tee-piece connection; 2 ball valve; 3 base nut; 4 piston;
5 blowdown valve body; 6 ball valve; 7 ferrule; 8 seating; 9 cap nut;
10 spring; 11 piston ring; 12 liner; 13 from boiler; 14 to drain;
15 steam supply from injectors

Left: A continuous blowdown valve fitted to the faceplate or backhead, where it draws the waste salts from the firebox crown. The small pipe leading to the le is the pipe from the Left hand injector, while the large vertical pipe at the front is the waste pipe to the ashpan. *Author*

working, admitting steam to lift a piston in the body of the blowdown valve. Connected to the piston is a push rod that lifts another ball valve, allowing the salts to escape from the boiler.

Not all locomotives are fitted with a continuous blowdown valve as its value depends on the type of water treatment used.

THE PASSAGE OF STEAM

Regulator valve

Steam generated in the boiler is collected in the dome or the highest part of the boiler; here is located the regulator valve or the steam collector pipe leading to the regulator valve in the smokebox. When the regulator handle on the footplate is moved it operates the regulator valve, which when opened allows steam to pass through to the main internal steam pipe. For more detail on regulator valve operation see Chapter 6.

The pipe takes the steam direct to the cylinder steamchests in a saturated-steam locomotive, while in a superheated locomotive the steam passes through the superheater first. In the steamchest, valves control the distribution of the steam through ports or apertures at each end of the cylinder. The valves also allow the exhaust steam to leave the cylinder through the same ports when it has done its work in the cylinder, whence it travels to the blastpipe and chimney.

Types of steamchest valves

There are three types of valves fitted in the steamchest of the locomotives in use today. The first is the **slide valve,** sometimes known as the 'D' valve, which is usually found on industrial locomotives and is always outside admission. Some slide valves work back-to-back in a common steam chest, while others are mounted in steamchests on top of the individual cylinders.

The second type is the **piston valve,** fitted to later main-line locomotives; this type can have either outside or inside admission. Piston valves are fitted with valve rings to keep them steamtight in the same way as the cylinder piston itself.

There are two piston valve heads per cylinder in the steamchest, working together on a common spindle known as the valve spindle, which enters the steamchest through a steamtight packing called the valve spindle gland. The front end of the steamchest is fitted with a steamtight cover that incorporates an extension for the tail rod of the valve spindle; the length of the spindle requires the use of a tail rod to keep the wear of the front valve to a minimum.

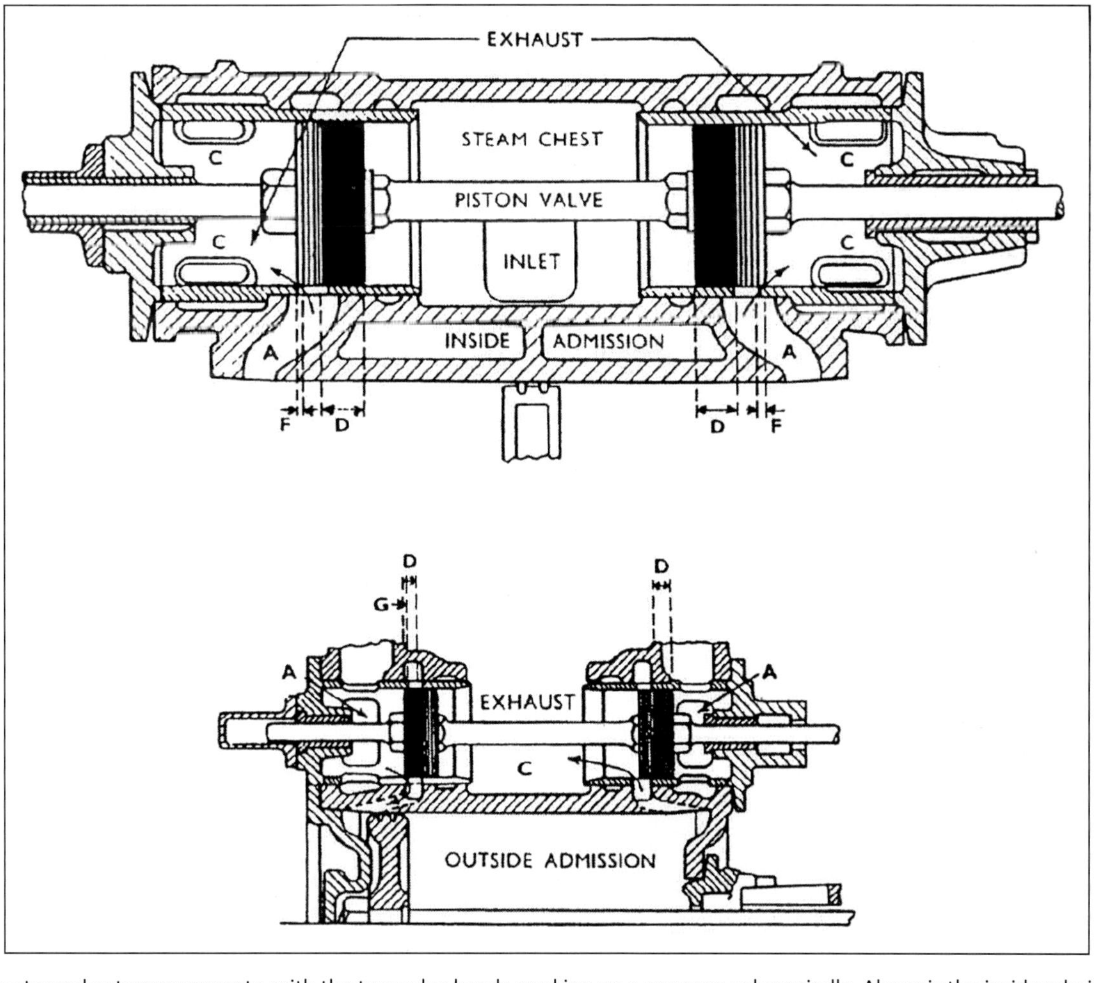

Piston valve steamchest arrangements, with the two valve heads working on a common valve spindle. Above is the inside admission design with the steam inlet between the heads. Below is the outside admission version, with the live steam contained outside the valve heads, but working on much the same principle: A steam port; B bridge; C exhaust port; D steam cap; E exhaust lap; F exhaust clearance; G lead; H angle of advance

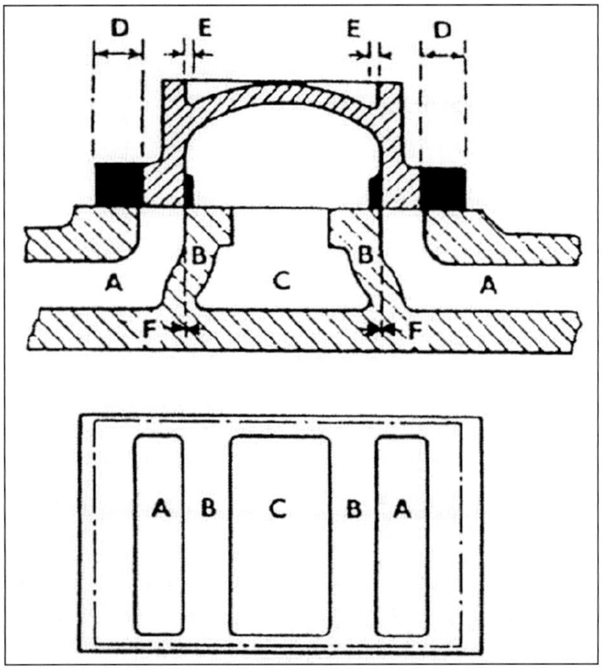

At each end of the steamchest there are two separate steamtight valve liners, each containing one piston valve head, with a space between them. With the inside admission design the live steam is admitted between the two valve heads, then after it has done its work in the cylinder it is exhausted via steam pipes at each end of the steamchest outside the valve heads. With outside admission the live steam is admitted outside the valve heads, while the exhaust aperture is between the two heads. (Outside admission is always used with slide valves, the steam pressure in the chest keeping the valve in contact with the face of the valve.)

The third type of valve is the **poppet valve,** as used with the Caprotti valve gear. These valves are operated by a camshaft in a similar way to the valves of an internal combustion engine, and are discussed in more detail below.

Lap and lead

'Lap' is the amount by which the valve overlaps each steam port at the middle position of each valve; 'steam lap' is the amount by which the valve overlaps the port on the live steam side, and 'exhaust lap' is the corresponding amount on the exhaust side. The latter is normally given to slow-running locomotives such as shunters, delaying the exhaust in order to derive the maximum work for the expanding steam in the cylinder.

The 'lead' of a valve is the amount by which the steam port is open when the piston is static at the front or back dead centre. Admitting steam at this point fills the space between the cylinder and piston and ensures maximum cylinder pressure at the beginning of the stroke. This is particularly important for high-speed locomotives, where the valve events are happening in rapid succession.

CYLINDERS

Steam locomotives are fitted with two, three or four cylinders, within which steam is alternately admitted and exhausted from either side of the piston. Unlike an internal combustion engine, the steam locomotive employs both sides of the piston. The cylinders have steamtight covers, and the piston rod passes through the rear cover by means of a steamtight packing called a piston gland. This can either be a soft packing

Above: Cross-section and plan of the steamchest arrangements for the slide or 'D' valve. The live steam is outside the valve and is admitted to the cylinders through steam port A and exhausted through port C: A steamport; B bridge; C exhaust port; D steam cap; E exhaust lap; F exhaust clearance

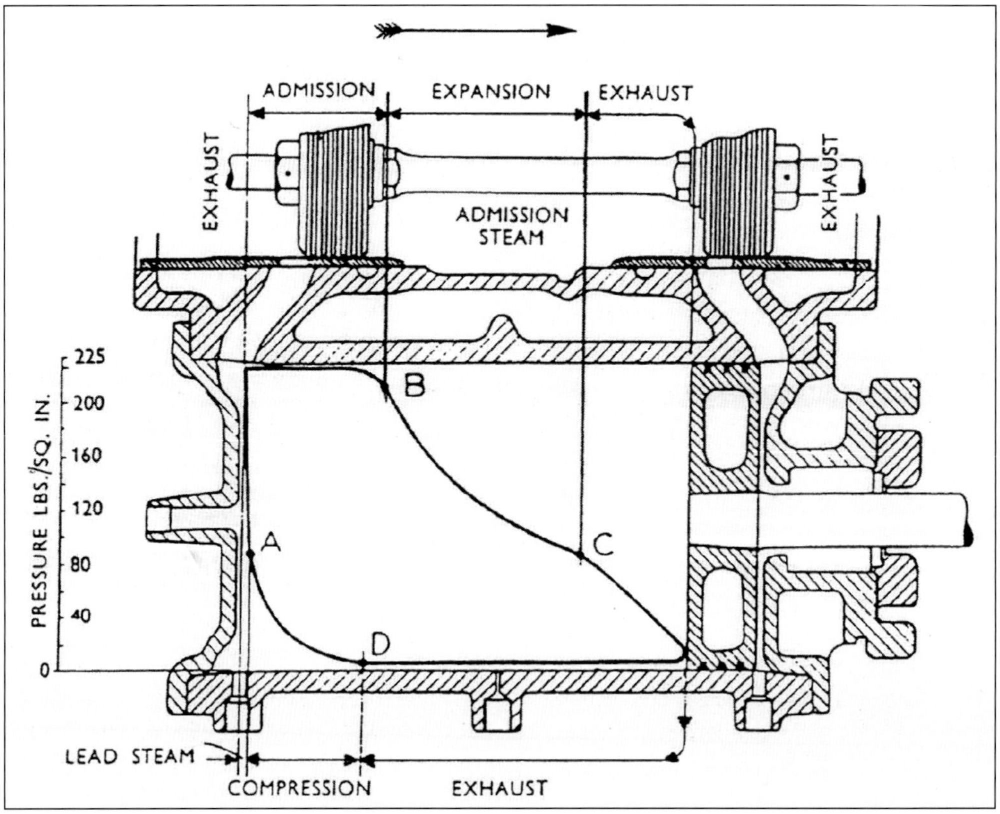

Right: Cross-section of a steam cylinder with inside admission, showing the percentage of valve travel or 'cut-off' in relation to the steam pressure in the cylinder: A point of admission; B point of cut-off; C point of release; D point of compression

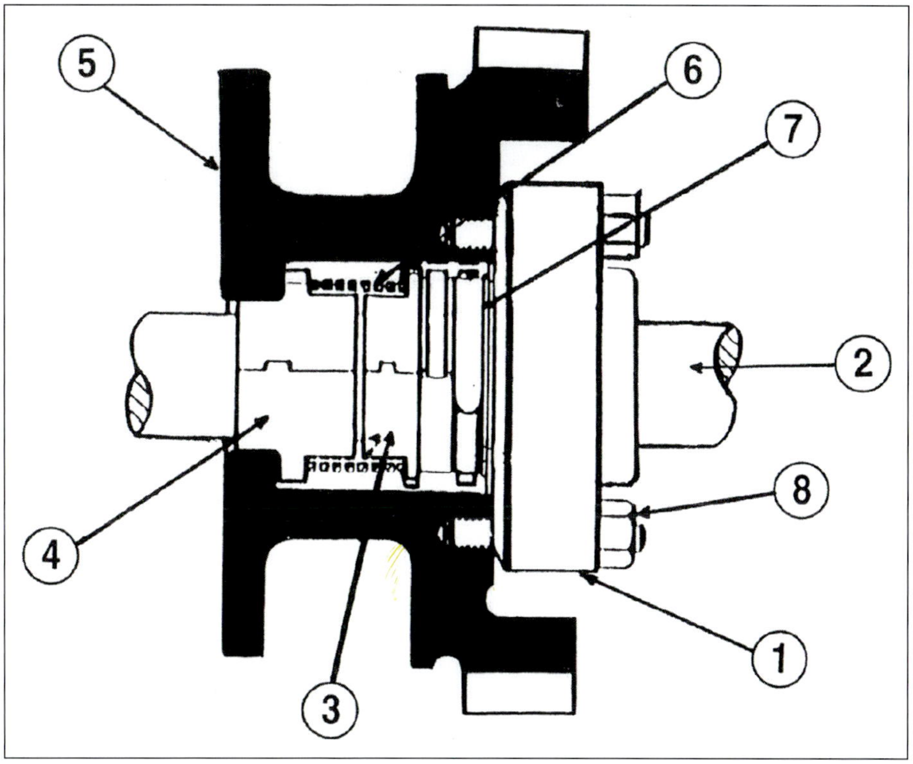

Above left: A split metal cylinder gland shown separated into the two halves that fit round the piston rod, retained by a spring clip; some of the clips are coil springs, while others are as shown. The second photograph shows the gland assembled for fitting round the piston rod. *Author*

Above right: Section of the spring-type piston gland: 1 outer housing; 2 piston rod; 3 outer gland packing; 4 inner gland packing; 5 cylinder end cover; 6 tension spring; 7 packing; 8 nut

Below left: The piston gland. Note the oil feed at the top, while just to the right of the gland can be seen the water relief valve. *Author*

Below right: The automatic water relief valve fitted to the cylinder end cover of an Ivatt Class 2 2-6-0. Just below can be seen the pipes of the cylinder drain cocks that are operated by the driver from the footplate. *Brian Dobbs*

Steamchest details showing the various valve positions during one revolution of the wheel for an inside admission piston valve, and an outside admission piston valve or slide valve. For Walschaerts or Stephenson link motion the direction of the valve with inside admission primarily follows the direction of the piston up to the end of the stroke; when the piston stops at wheel dead centre, the valve continues its movement due to the action of the union link and combination lever. With outside admission the valve moves mainly in the opposite direction to the piston until the end of the stroke where, as with inside admission, the valve continues to move.

compressed into a gland by a sleeve around the piston rod, or two split circular sections made from cast iron, which is kept steamtight against the piston rod by coil tension springs. The piston itself is kept steamtight in the cylinder by piston rings made from cast iron, similar to those fitted to the piston of an internal combustion engine. Due to the weight of the piston working in a horizontal cylinder, some are fitted with springs on the lower quarter between the rings and the piston itself, helping to carry the weight and prevent the piston from wearing to an oval shape.

Fitted to both the cylinder end covers are automatic water pressure relief valves that open when the pressure in the cylinders reaches about 2016 above boiler pressure. If water is allowed to build up in the cylinders and the pressure is not relieved by the driver opening the cylinder drain cocks from the footplate, the relief valves open to prevent hydraulic pressure from blowing out the end covers.

The work of the steam in the cylinder

The following is a description of the sequence of events during one revolution of the driving wheel, starting with the piston at the leading end of the cylinder.

Steam enters the cylinder at a little less than boiler pressure (due to steam losses) through the port at the front, and forces the piston backwards along its stroke or sweep. With the locomotive in full forward gear the steam will be admitted to the cylinder and provide power for the piston for about 75 per cent of its travel. During this stroke the valve spindle is also moving until the valve heads close the admission ports. The steam supply is then cut off and the steam expands until the end of the stroke.

This point is called 'dead centre', which means that although the piston is stationary at the end of its stroke, the wheel continues to revolve. The piston now reverses its direction and begins to move forward. The valve also changes its direction and moves to open the exhaust port to allow the steam to escape from the cylinder ahead of the piston, again for about 75 per cent of its stroke; meanwhile steam is admitted behind the piston. The exhaust steam passes to the blastpipe until the exhaust valve closes. Then the exhaust steam left in the cylinder is compressed by the forward movement of the piston (see 'Lap and lead' above).

The piston is connected to the piston rod which passes through the piston gland and is connected to the crosshead. This is made from forged steel with a white metal surface, and runs between slide bars that guide it in its backward and forward movement and prevent the piston rod from bending. Also connected to the crosshead is the connecting rod, its little end bearing pivoting on what is called the gudgeon pin. At the opposite end of the connecting rod is the big end on the driving wheel crank; the pivoting action of the big and little ends of the connecting rod converts the reciprocating action of the piston into the revolving motion of the driving wheels.

VALVE GEAR

The drive for the steamchest valves is derived from the locomotive's driving wheels, and several different designs of valve gear are in common use.

Walschaerts valve gear

Versions of the Walschaerts motion can be used for slide valves or inside or outside admission piston valves, and can also be used inside or outside the locomotive frames.

The Walschaerts motion derives its drive from two independent sources brought together in the valve spindle

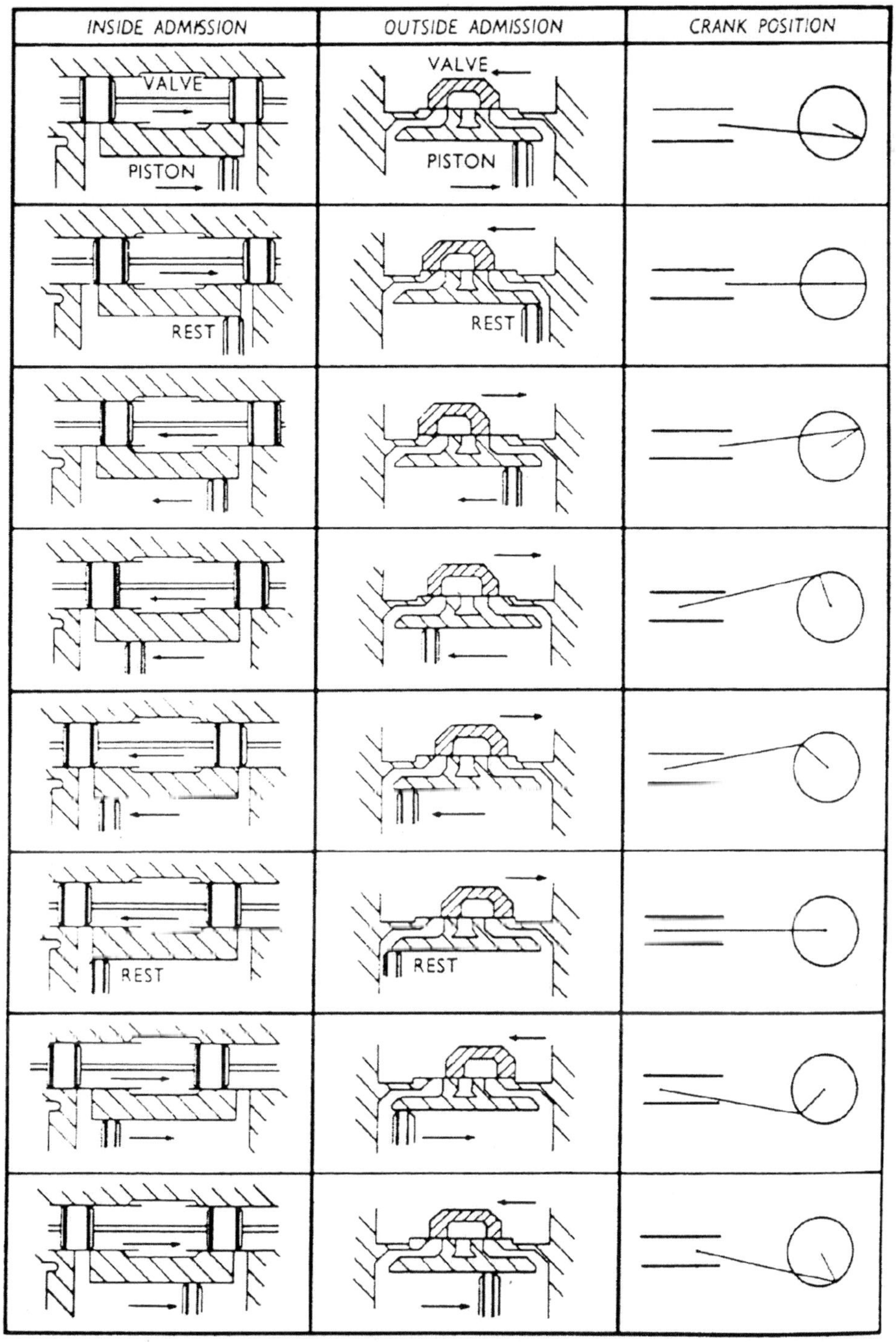

The Midland Railway-pattern water relief valve fitted to the cylinder casting instead of the end cover. *Author*

crosshead. The first is the return crank rod (outside motion), or the eccentric rod (inside motion) fitted to the driving axle. The return rod converts the radial motion of the crank into the reciprocating action necessary for operation of the valves. The return rod is connected to the end of the expansion link, which it causes to rock to and fro. Driven from a die block in the expansion link is the radius rod; the position of the die block in the link controls the steam cut-off for the valves and the direction of the locomotive.

The other drive source comes from the combination lever, sometimes called the pendulum lever, attached via the union link to the crosshead; this provides the additional movement of the valve to provide lead steam (that which is admitted just before the end of the stroke of the piston).

With inside admission valves the eccentric or return crank is set at 90 degrees behind the crank, while with outside admission it is 90 degrees in front of the crank; this is so the valve travels in the required direction. It will be noted from the diagram of valve positions (above) that with inside admission the valve primarily travels in the same direction as the piston, whereas with outside admission it travels in the opposite direction from the piston.

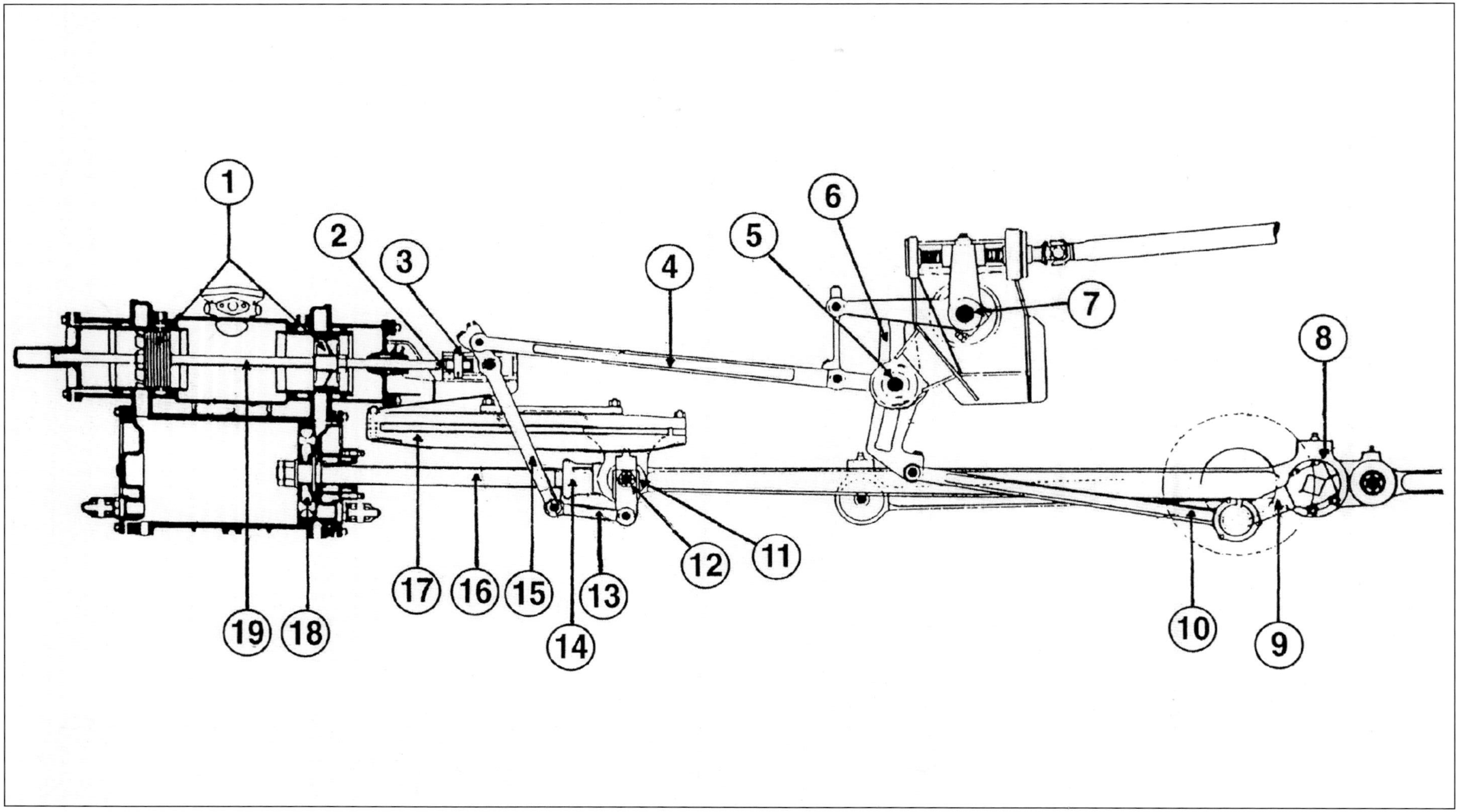

lvi) Diagram of an inside admission Walschaerts valve gear: 1 piston valve heads; 2 valve spindle crosshead guide; 3 valve spindle crosshead; 4 radius rod; 5 expansion pivot with die block inside expansion link; 6 expansion link; 7 reversing shafts; 8 connecting rod big end; 9 eccentric (inside motion)/return crank (outside motion); 10 eccentric rod (inside motion)/return crank rod (outside motion); 11 connecting rod small end; 12 gudgeon pin; 13 union link; 14 crosshead; 15 combination lever; 16 piston rod; 17 slide bar; 18 piston head; 19 valve spindle

A view of a typical Walschaerts valve gear, as fitted to British Railways Standard locomotive No 76079. *Brian Dobbs*

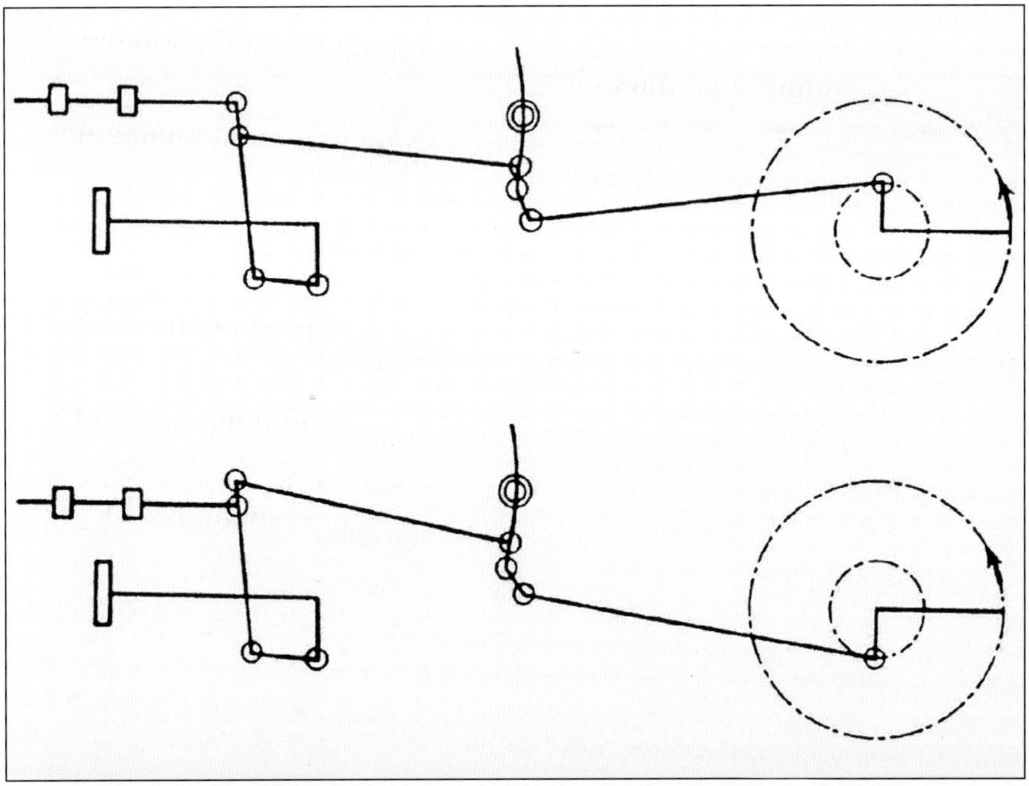

Left: Diagrams of the Walschaerts motion showing the difference between outside (above) and inside (below) admission, in particular the pivot between the radius rod and combination lever.

Below: Outside admission Walschaerts valve gear fitted to a Somerset & Dorset 2-8-0; note that the valve spindle is above the radius rod/combination lever pivot. The radius rod and expansion link are connected to the expansion die-block, and can be lifted in the die-block to alter the direction of the locomotive. *Mrs. Wendy Smith*

A photograph showing the return crank rod connected to the radius rod via the expansion link. Also shown is the union link from the crosshead connected to the combination lever. *Author*

When the piston is at rest, the crosshead and therefore the combination lever are also at rest, but as the return crank is set out of phase with the crank pin, it gives continuous movement to the valve spindle via the radius rod. It may also be noted that where the locomotive uses outside admission the valve spindle is above the pivot with the radius rod and combination lever, which alters the direction of the valve to the opposite of that of the piston, because the live steam is outside the two valve heads and therefore the exhaust is in between them. With inside admission, the valve spindle is below the pivot point.

Stephenson link motion

Another common type of valve gear is the Stephenson link motion. This utilises two eccentrics per cylinder, both connected by eccentric rods to the same expansion link; one is connected to the bottom of the link and controls the backward direction of the locomotive, while the other is connected to the top to control the forward direction. From the expansion die-block the travel of the valve is transmitted to the valve spindle by the intermediate valve spindle running through the motion bracket.

Like the Walschaerts valve gear, the Stephenson link motion can be used for inside or outside admission and has the benefit of increasing lead steam as the cut-off is reduced from full gear.

Bulleid chain-drive valve gear

On certain of Bulleid's locomotives for the Southern Railway a chain drive is used to operate the piston valves instead of return cranks or eccentrics. The drive is taken from the centre driving axle of the locomotive to an intermediate chain wheel, from where it is taken again by chain to a three-throw jockey shaft. A link mechanism joins the jockey shaft to the expansion links that control the cut-off for the valves.

For ease of operation by the driver, the reversing gear and cut-off for the locomotive is operated from the footplate by a double steam cylinder connected to the equivalent of the radius rod by a link mechanism.

Caprotti valve gear

The last type of valve gear to be considered is the British Caprotti poppet valve gear, which is suitable for two-, three- or four-cylinder locomotives. In its three-cylinder form the valves are operated by cams in camboxes driven from shafts connected to worm drives on the return cranks of the centre driving wheels. Short rear driving shafts with universal joints at each end allow the wheelset to slide up and down within the axlebox horn guides. Fitted to the top of the worm drive box on the return crank is an anchor link to the anchor bracket on the main frame of the locomotive. An intermediate driving shaft transmits the power via a short front driving shaft to the cambox on top of each of the locomotive's outside cylinders. From there the drive is taken by a cross shaft to the inside cylinder.

The valves in the steamchest are driven by cams; each cylinder has four valves, two inlet and two exhaust.

The two-cylinder arrangement is similar except that the drive is taken from one common drive box with a worm gear fitted to the leading driving wheel axle, then through a system of shafts and a cross shaft to the cam boxes above each cylinder steamchest.

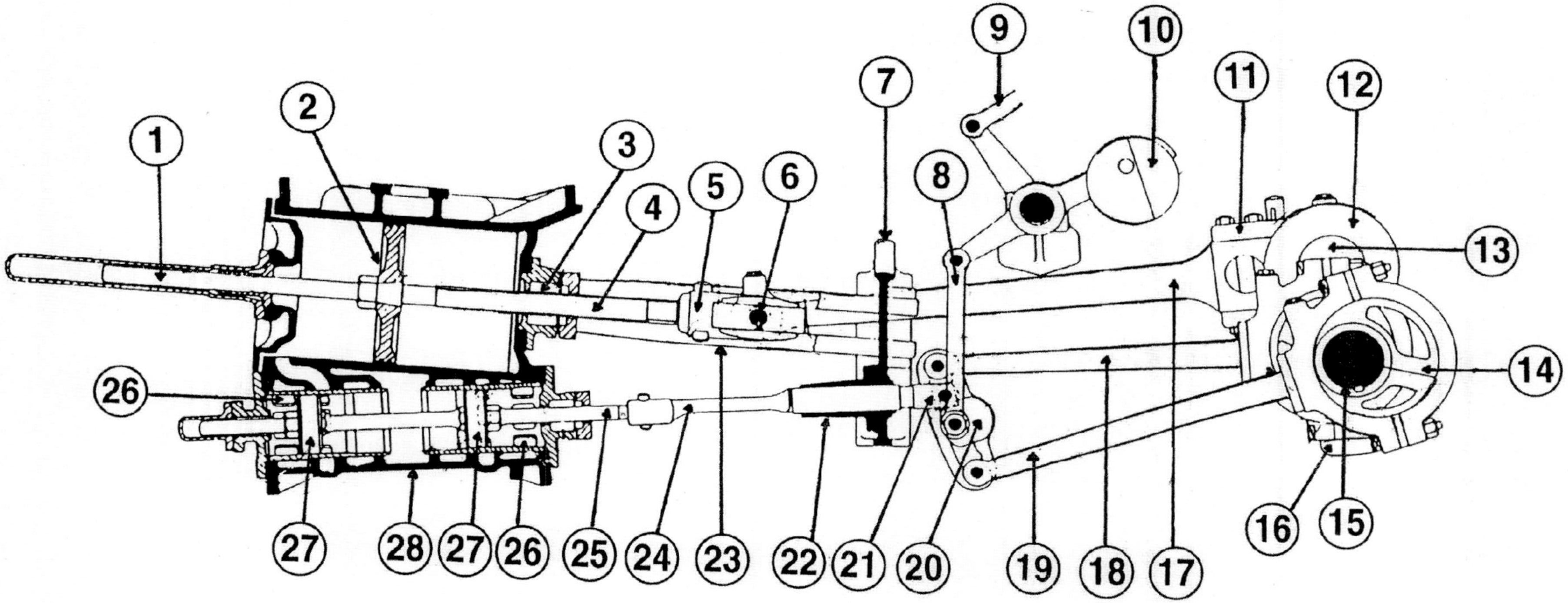

Diagram of the Stephenson link motion valve gear, with inside motion and piston valves: 1 tail rod; 2 piston head; 3 packing and gland; 4 piston rod; 5 crosshead; 6 gudgeon pin; 7 motion plate; 8 lifting link; 9 reversing or reach rod; 10 balance weight; 11 big end strap; 12 crank web; 13 crank pin; 14 eccentric sheaves; 15 crank axle (built up); 16 crank web balance extension; 17 connecting rod; 18 forward eccentric rod; 19 backward eccentric rod; 20 expansion link; 21 die black; 22 valve spindle guide; 23 slide bars; 24 intermediate valve spindle; 25 valve spindle; 26 steam ports; 27 piston valve heads; 28 steamchest

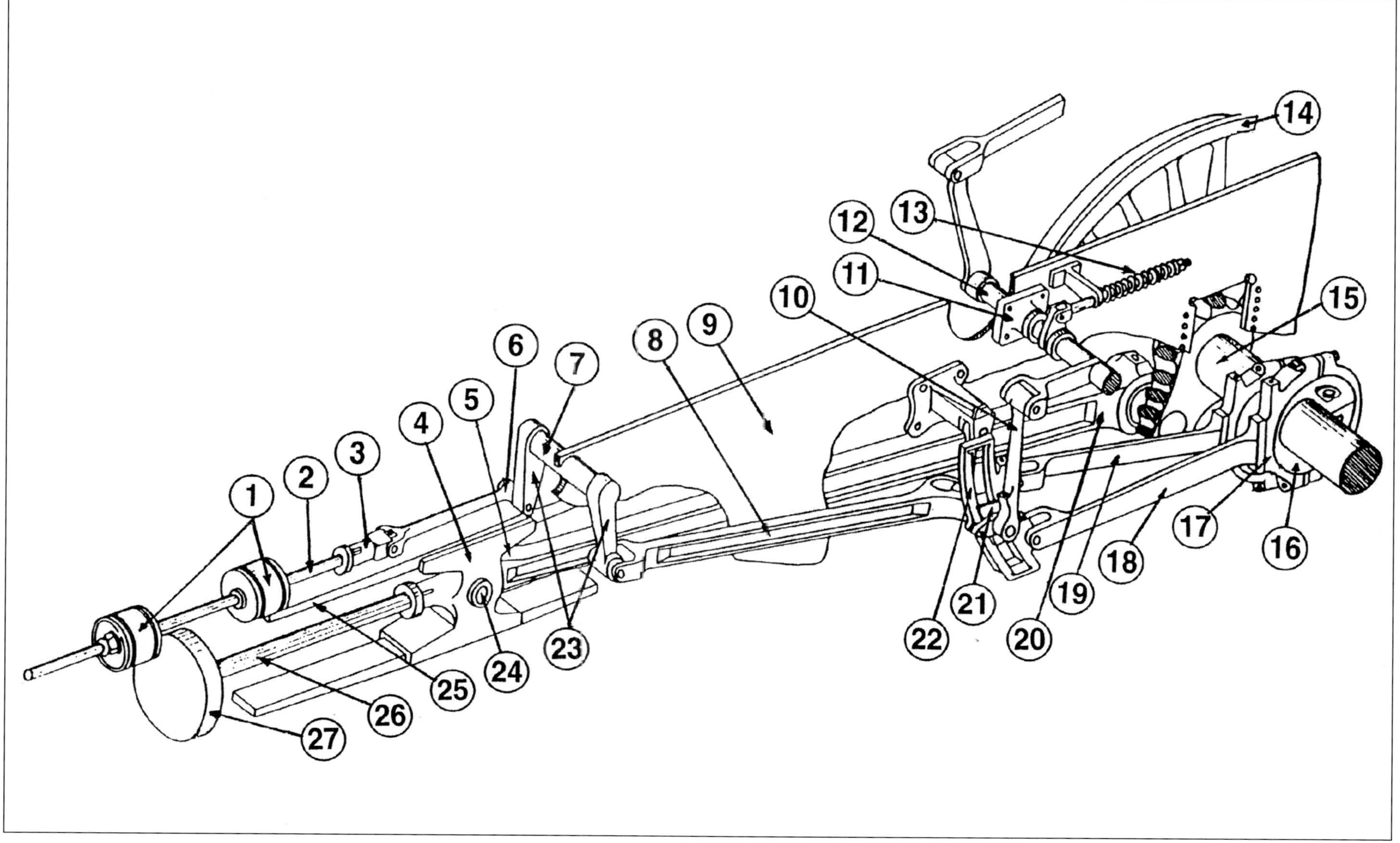

Another version of the Stephenson link motion, with outside cylinders and inside motion. This type of arrangement can often be found on former Great Western locomotives: 1 piston valve heads; 2 valve spindle; 3 valve spindle crosshead; 4 crosshead; 5 connecting rod small end; 6 valve link; 7 rocking shaft; 8 intermediate valve rod; 9 mainframe; 10 lifting links; 11 reversing shaft bracket; 12 reversing shaft; 13 reversing shaft spring; 14 driving wheel; 15 axle; 16 eccentric sheave; 17 eccentric strap; 18 backward eccentric rod; 19 forward eccentric rod; 20 connecting rod big end; 21 die block; 22 expansion link; 23 valve arms; 24 gudgeon pin; 25 slide bars; 26 piston rod; 27 piston head

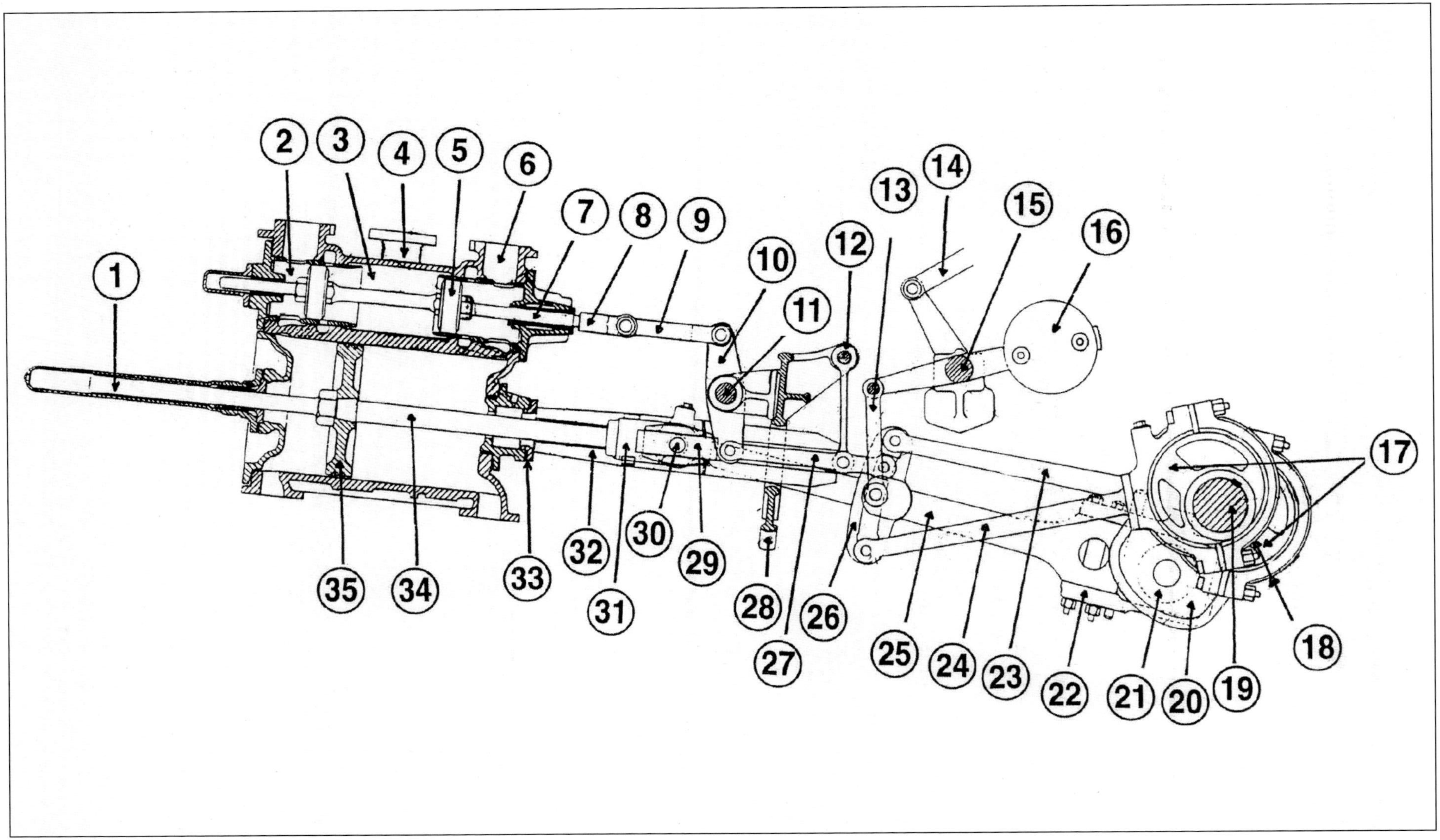

A third example of the Stephenson link motion, where the drive to the valves is taken through rocking levers to operate the valves above the cylinder: 1 tail rod; 2 exhaust port; 3 steamchest; 4 steam pipe; 5 piston valve head; 6 exhaust port; 7 valve spindle; 8 valve spindle crosshead; 9 connecting link; 10 rocking lever; 11 rocking shaft and bracket; 12 swing link; 13 lifting link; 14 reversing rod; 15 reversing shaft and bracket; 16 balance weight; 17 eccentric sheaves; 18 eccentric straps; 19 crank axle; 20 crank web; 21 crank pin; 22 big end strap; 23 forward eccentric rod; 24 backward eccentric rod; 25 connecting rod; 26 expansion link; 27 intermediate valve rod; 28 motion plate; 29 slide block; 30 gudgeon pin; 31 crosshead; 32 slide bars; 3 packing and gland; 34 piston rod; 35 piston head

Ex-LMS 'Black Five' No 4767, named George Stephenson, showing the Stephenson link motion. *Brian Dobbs*

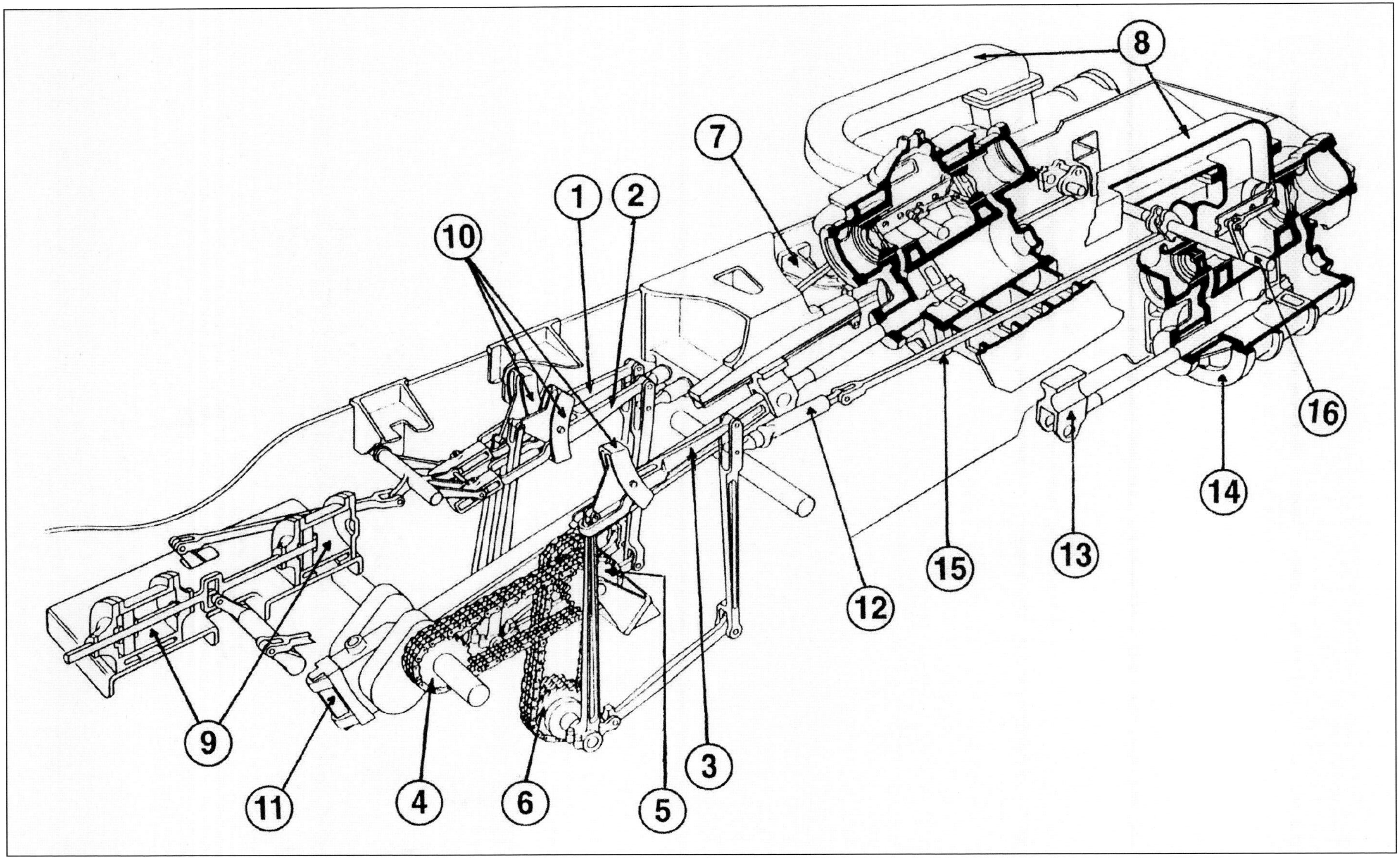

The Bulleid chain-drive valve gear fitted to the three-cylinder Southern Railway 'West Country' class 'Pacific' locomotives. Note the arrangement of the cylinders, with the chain drive for the valve gear taken from the centre driving axle to the jockey shaft, which operates the expansion links that control the valves: 1 extreme left-hand valve gear for middle (inside) cylinder; 2 second set of valve gear for left-hand outside cylinder; 3 third set of valve gear for right-hand outside cylinder; 4 chain wheel on centre driving axle; 5 intermediate chain wheels on bracket; 6 chain wheel driving three-throw jockey shaft; 7 hanging bracket to offset second valve drive; 8 left and right outside cylinder exhaust led back to exhaust of inside cylinder; 9 steam reversing cylinders; 10 expansion links; 11 inside cylinder big end; 12 valve rod extension; 13 outside (right-hand) cylinder crosshead; 14 outside (right-hand) cylinder; inside cylinder; 16 connection to valve spindle

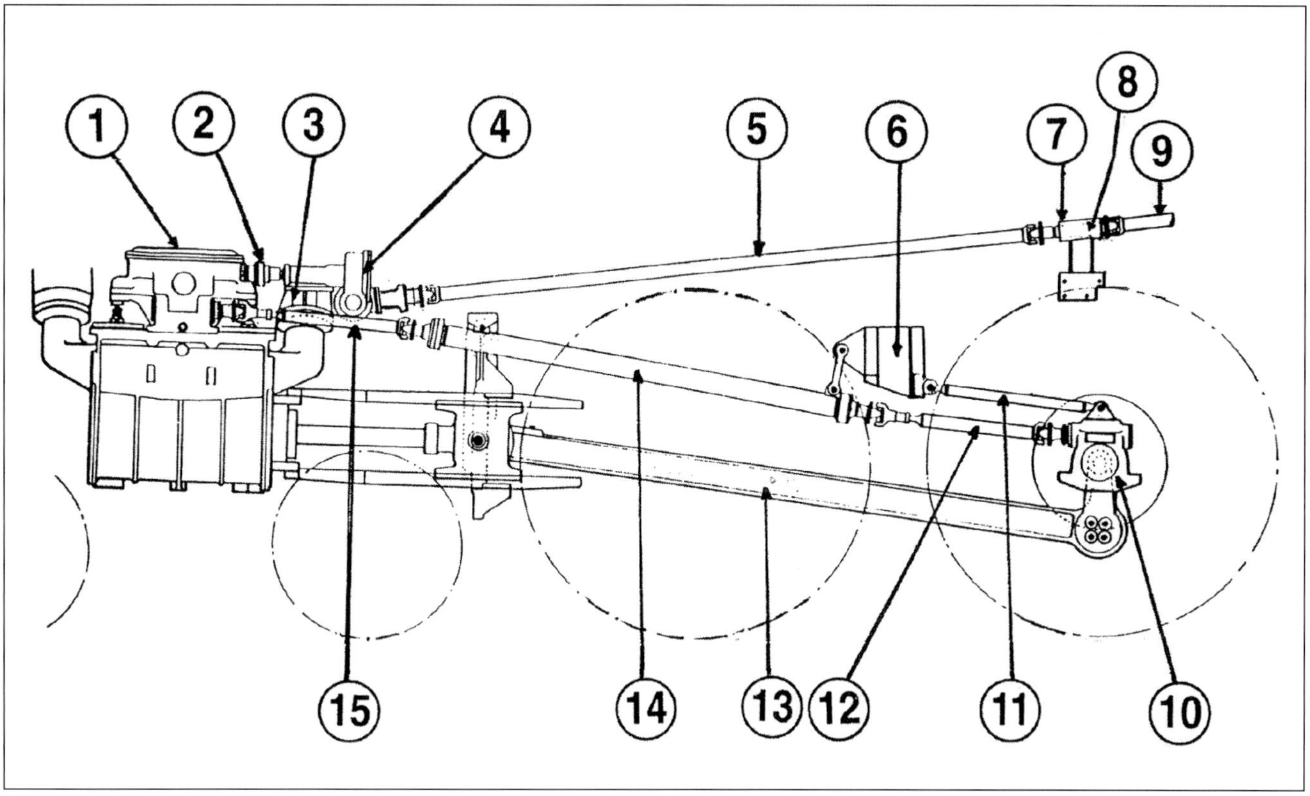

Diagram of the British Caprotti valve gear, for outside drive: 1 cambox; 2 cambox coupling; 3 carrier; 4 worm and bevel gearbox; 5 front reversing shaft; 6 anchor link bracket; 7 intermediate bearing; 8 intermediate shaft; 9 back reversing shaft; 10 return crank gearbox; 11 anchor link; 12 rear driving shaft; 13 connecting rod; 14 intermediate driving shaft; 15 front driving shaft

The general arrangement of the Caprotti valve gear on the right-hand side of the locomotive, with the drive from the centre driving wheel to the cambox. *D. Dyson*

Left and below: This is the left-hand side of the locomotive, and shows the worm drive gearbox of the Caprotti gear connected to the return crank, and the anchor link. The second photograph, also from the left, shows the end cover of the gear that takes the power to the cross shaft on the inside cylinder valve gear. *D. Dyson*

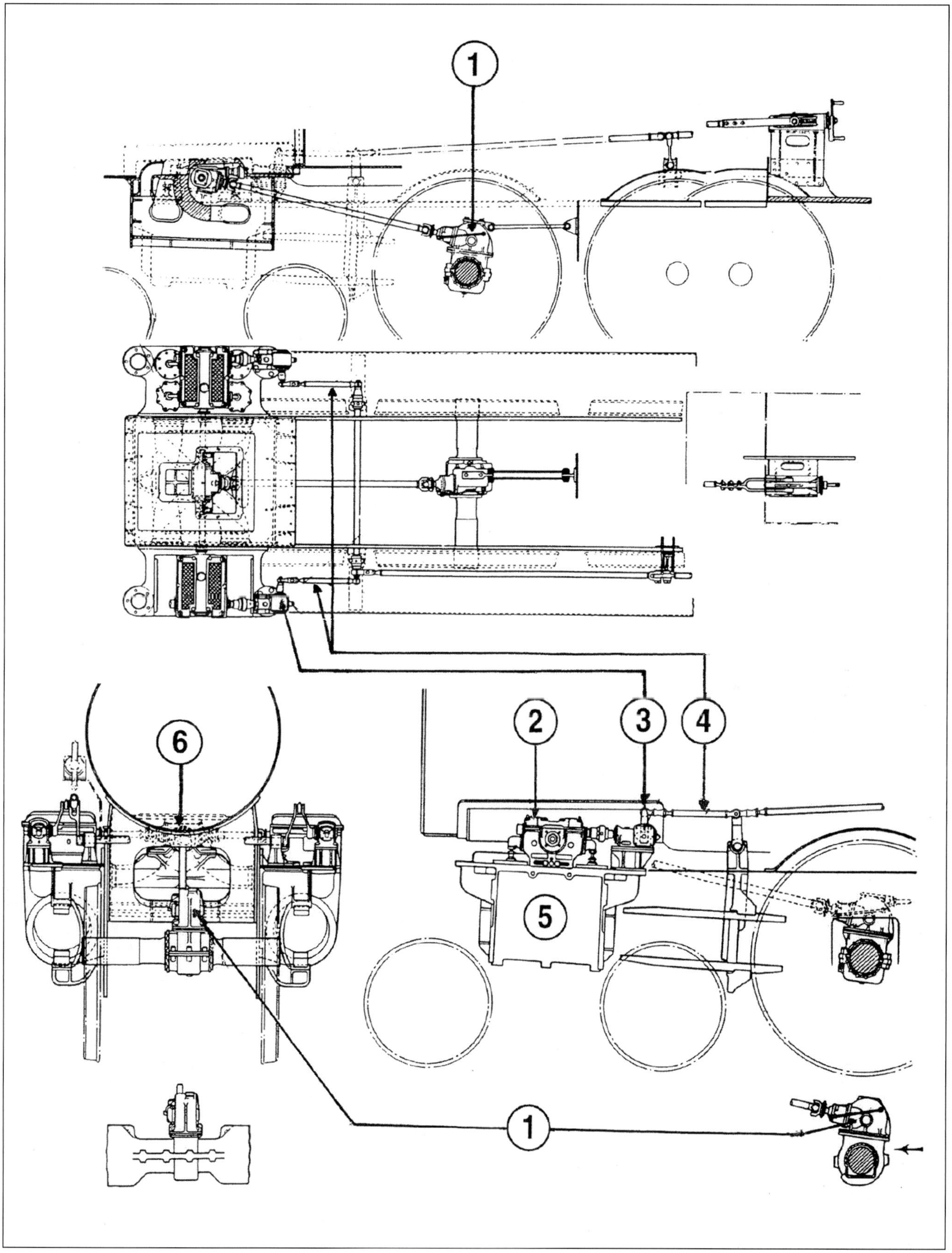

Diagrams of the Caprotti valve gear drive taken from the inside leading axle of a two-cylinder locomotive, as opposed to the outside driving axles. The upper diagrams show the plan and elevation, the lower diagrams the end view (as seen from the rear of the engine) and the outside side view: 1 drive from axle; 2 cambox; 3 reversing gear box; 4 reversing rod; 5 cylinder; 6 cross driving gear

Joy valve gear has not been included as there are very few locomotives in steam on operating railways with this gear.

Reversing

When the direction of a locomotive needs to be changed from forward to reverse, the valves must be moved in the steamchest to change the admission of steam from one side of the piston to the other. This is done by moving the die-block in the expansion link, using a lifting link connected to the radius rod. As the reversing mechanism is operated from the footplate and the radius rod is lifted or lowered, the expansion link pivots in the motion and the die-block is moved in the expansion link, causing the valve to move in the steamchest and alter the steam admission. If the die-block is in the bottom half of the expansion link the locomotive will be in forward gear; if it is in the top half it will be in reverse gear.

Valve control on multi cylinder locomotives

There are various designs of motion to operate the inside steamchest valves in three- or four-cylinder locomotives.

It was common to have the valves of the inside cylinder (or what is sometimes called the inside engine) operated by a conjugated linkage system taken from the outside cylinder, as is the case on former LNER 'Pacifies' such as *Flying Scotsman.*

The LMS, meanwhile, adopted three independent sets of motion for its three-cylinder 4-6-0 express locomotives, with the valve on the inside cylinder being controlled by an eccentric. On the four-cylinder LMS 'Pacifies' the valves for the inside cylinders were controlled by a pivoting rocking shaft taken from the outside cylinders.

The Great Western Railway also adopted rocking shafts for its four cylinder 'Castle' and 'King' class locomotives, driven from the Walschaerts motion of the inside cylinders.

Compound locomotives

The development of 'compounding' arose from a desire to save coal by making the steam locomotive more efficient, even at a time when fuel was comparatively cheap.

In a compound engine, instead of the spent steam being exhausted from the cylinders to the atmosphere, it is re-used in low-pressure cylinders, therefore giving the locomotive greater economy of coal and water. In practice, compounding offered little actual economy over a well-designed 'simple' locomotive. It was also complicated in operation, so was little used in the UK.

When a compound locomotive moves away from a start or is running at low speeds it operates as a simple engine, with the steam being admitted to both high- and low pressure cylinders. As the regulator handle on the footplate is moved over and speed increases, it causes admission of live steam to

The right-hand outside cylinder of a former Great Western Railway locomotive, showing the valves operated by a rocking shaft from the Walschaerts motion of the inside cylinders. *Peter Herring*

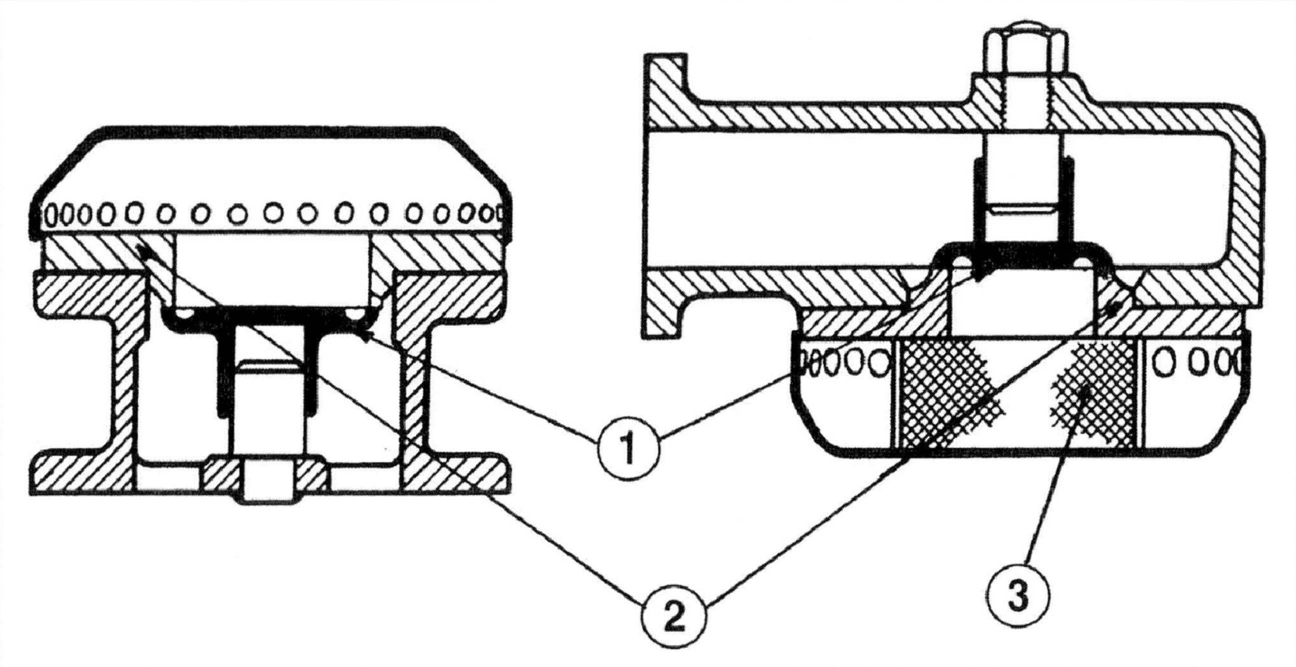

the low pressure cylinder to be shut off, and 'compounding' takes place. The exhaust steam from the high-pressure cylinder now passes to the low pressure cylinder before being exhausted through the blastpipe, thus working twice.

Above: Cross-sections of anti-vacuum valves fitted to Standard Class 4 locomotives. These are similar to most examples either attached to the steamchest or superheater header: 1 valve; 2 removable seating; 3 sieve

Anti-vacuum valves

Anti-vacuum valves, or what are sometimes called 'snifting valves', are fitted to locomotives with piston valves, either on the superheater or on the steamchest. Their purpose is to admit air automatically to the steamchest or superheater and destroy the vacuum caused by the action of the pistons and valves when the locomotive is coasting with the regulator closed. The vacuum created in the cylinders might otherwise draw ash and carbon into them from the smokebox.

The valve is kept on its seat by steam pressure when the regulator is open; when it is closed the valve falls open by gravity. When the regulator is re-opened a puff of steam will be seen to pass the valve and escape into the atmosphere as the steam pressure forces the valve back on to its seat. The air can be heard as a jingling noise passing through the valve when the locomotive is coasting.

Anti-vacuum valves are not necessary on slide valve locomotives as they are all outside admission, so when the locomotive is coasting there is no steam pressure in the steamchest to keep the valve on the valve face, so no vacuum is created in the cylinders.

LUBRICATION

Lubrication of a steam locomotive is achieved by several methods.

Hydrostatic lubricator

The hydrostatic lubricator provides lubrication for the cylinders, valves and pistons. The oil is carried through oil pipes as a solid stream and is mixed with steam from the

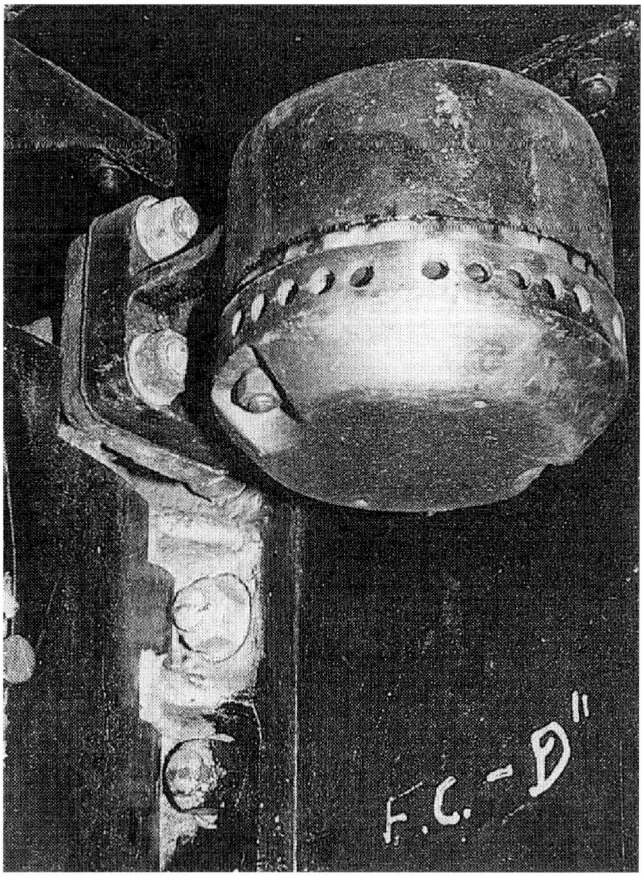

Above: The anti-vacuum valve fitted to the steamchest of a 'Black Five' 4-6-0. It cannot be easily seen as it is concealed under the running plate of the locomotive. *Author*

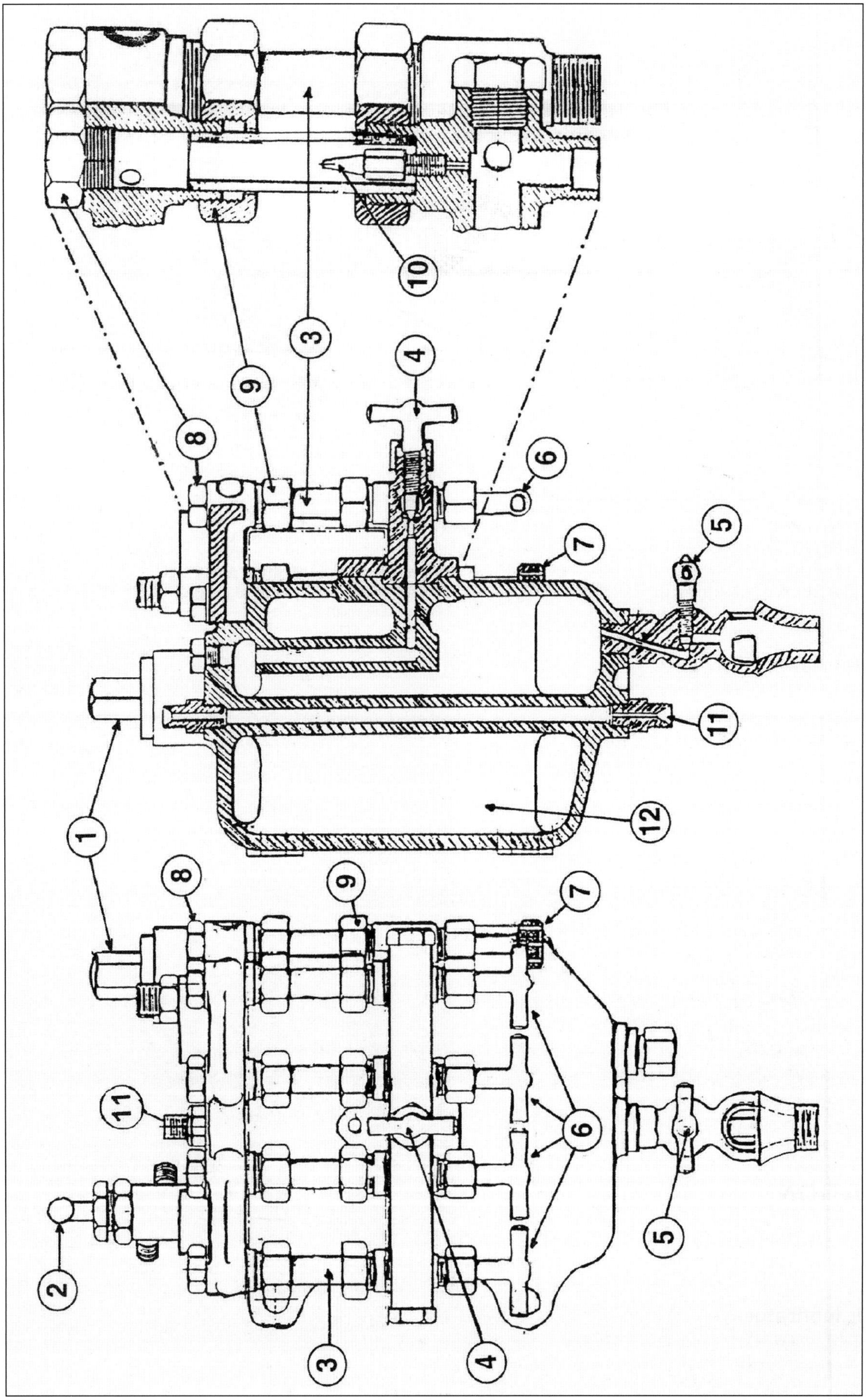

Front and side views of a four-feed hydrostatic lubricator as fitted to a Great Western locomotive: 1 oil filler cap; 2 steam cock; 3 sight glasses; 4 main gallery shut-off; 5 drain cock; 6 regulator valves; 7 regulator valve; 8 cap nut

A typical Great Western hydrostatic lubricator fitted to a pannier tank locomotive. As can be seen, this type has three sight-glasses. Fitted to the bottom is the drain cock with the drain. At the front of the lubricator oil gallery is the shut-off valve and the three regulator controls. On top of the body on the left is the steam cock, while the domed nut on the right is the oi filler. *Brian Dobbs*

A displacement lubricator fitted to former Southern Railway 4-6-0 No 30777 Sir Lamiel. It has two sight glasses, and the filler is at the top with the square-shaped nut. At the bottom of the reservoir is a drain nut to empty water from the body, while at the bottom of the sight feed glasses is the regulator to adjust the flow of oil. *Brian Dobbs*

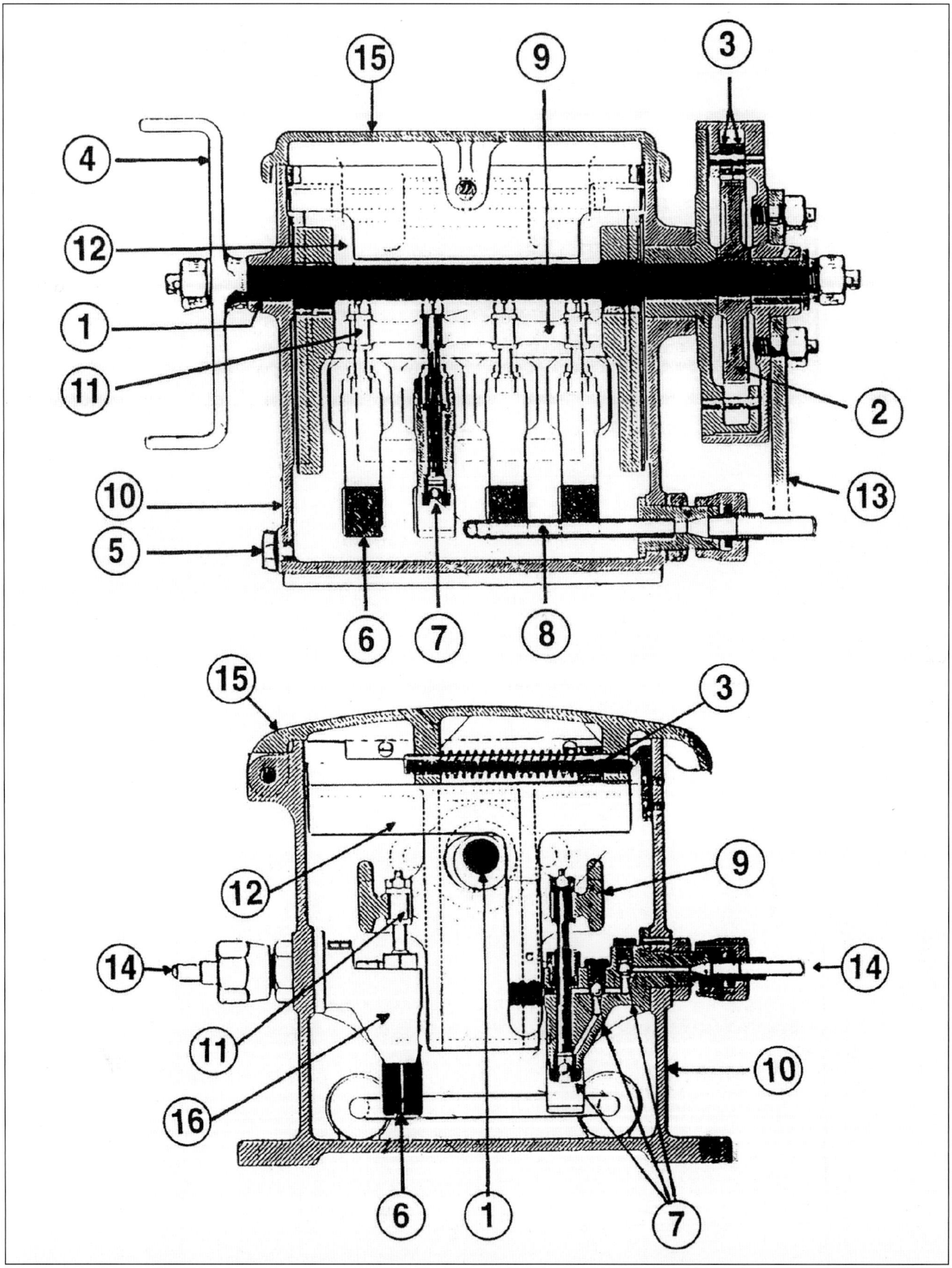

Side and end cross-sections of a typical mechanical lubricator: 1 camshaft; 2 ratchet mechanism; 3 sprung pawls; 4 handle for priming; 5 drain plug; 6 sieve; 7 ball valves; 8 warming device; 9 actuating frame; 10 lubricator body; 11 'thimble'; 12 top sieve; 13 operating lever from motion; 14 oil feeds; 15 lid; 16 pump body

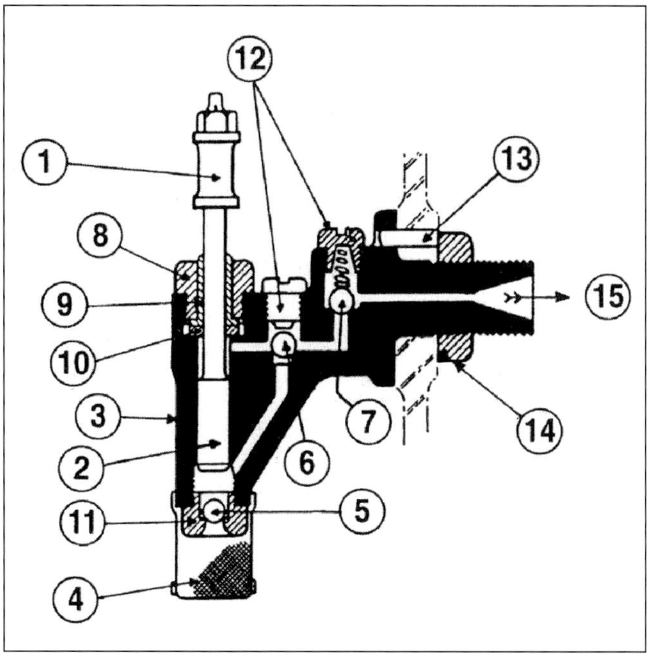

Left: Detailed diagram of a mechanical lubricator pump: 1 'thimble' with slot to locate with driving frame; 2 plunger; 3 body of pump; 4 sieve; 5 suction ball valve; 6 ball valve in cavity; 7 non-return ball valve; 8 nut; 9 packing; 10 seal; 11 base nut with seating; 12 sealing nuts; 13 body of lubricator; 14 securing nut; 15 oil delivery

Below: The atomiser, which heats the oil and mixes it with the steam, changing its consistency to the creamy liquid that is fed to the cylinders and steamchests: 1 oil from lubricator; 2 ball valve; 3 mixing chamber; 4 delivery; 5 steam supply; 6 spring

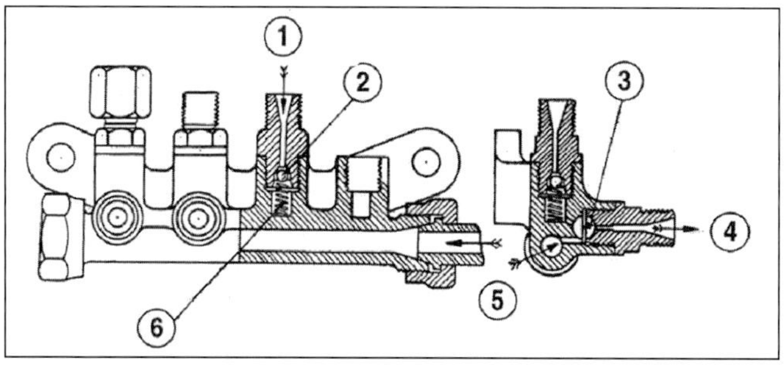

boiler in the atomiser. The steam condenses to very hot water, which emulsifies with the oil to form a creamy liquid. This is then fed along the main internal pipes into the steam chest and cylinders.

The hydrostatic lubricator can be found on most former Great Western locomotives, as well as others. The GWR examples have two to four sight-glasses and adjusting regulator valves, together with a separate assembly on the side of the lubricator body feeding directly to the regulator valve in the smokebox.

Mechanical lubricator

The mechanical lubricator is driven from a convenient part of the motion of the locomotive, and thus only operates when the locomotive is moving. As with the hydrostatic lubricator, the oil has to be emulsified before being fed to the cylinders. To achieve this the mechanical lubricator is fitted with a device known as an 'atomiser'; this heats the oil and mixes it with the steam, which changes its consistency to the creamy liquid that is fed to the cylinders and steamchests. Both hydrostatic and mechanical lubricators use only a special oil that is very thick in order to compensate for the heat given off by the steam in the cylinders and steamchest.

An axlebox lubricator fitted to the running plate of a locomotive. The pipes from the lubricator run to the axleboxes. At the end of the lubricator casing can be seen the ratchet that drives the lubricator from an arm connected to the motion. *D. Dyson*

Above left: The drive linkage to the axlebox and cylinder lubricator on an LNER A4 'Pacific' locomotive, taken from an arm on the trailing driving wheel. *Author*

Above right: An oilbox fitted to the framing of a locomotive; a worsted trimming drains the oil to the lubrication point. *Brian Dobbs*

A version of the mechanical lubricator can be used for the driving wheel axleboxes of the locomotive. These parts present no particular problem with heat, so normal lubricating oil can be fed directly from the lubricator to the point of lubrication.

Inside the mechanical lubricator, beneath the sieve, is a bank of pumps driven from a camshaft. There can be as many as 12 individual pumps in the lubricator depending on the number of cylinders fitted to the locomotive. The pumps are totally immersed in the oil in the lubricator body, which forms the reservoir, and are connected to the camshaft by an actuating frame that locates on 'thimbles' fitted to the shaft of each pump; the thimbles control the up and down movement of the plunger in the individual pump body. As the plunger moves up, oil is drawn through a sieve at the bottom of the body, then passes to a ball valve in the pump. At the end of the stroke the direction of the plunger is reversed, causing the ball to fall back on to its seating. The pressure keeps the valve closed, forcing the oil along a passage to a cavity containing another ball valve. The oil is now forced past a spring loaded non-return ball valve out of the pump and along copper pipes to flexible connections with the axlebox feeds; the pipes are flexible to allow for the up and down movement of the box in the frames. The feed pipes are connected to back-pressure valves containing further sieves and ball valves.

In the case of the cylinders, the oil will pass first to the atomisers to be turned into a creamy liquid before being fed through back-pressure valves to the cylinders. The steam supply to the atomisers is taken from a connection on the side of the smokebox where there is a 'tell-tale' valve, so called because when it is closed or partly closed steam can be seen emitting from the centre of the valve. The steam supply to the atomisers is also controlled from the cylinder drain cocks by a linkage, so that when the locomotive is standing with the drain cocks closed, the steam supply to the atomisers is closed off.

Another version of axlebox lubricator is the 'fountain' type, in which the oil is fed to the axleboxes under pressure from a reservoir.

Drip-feed lubricator

The third method of lubrication is by drip feed to various parts

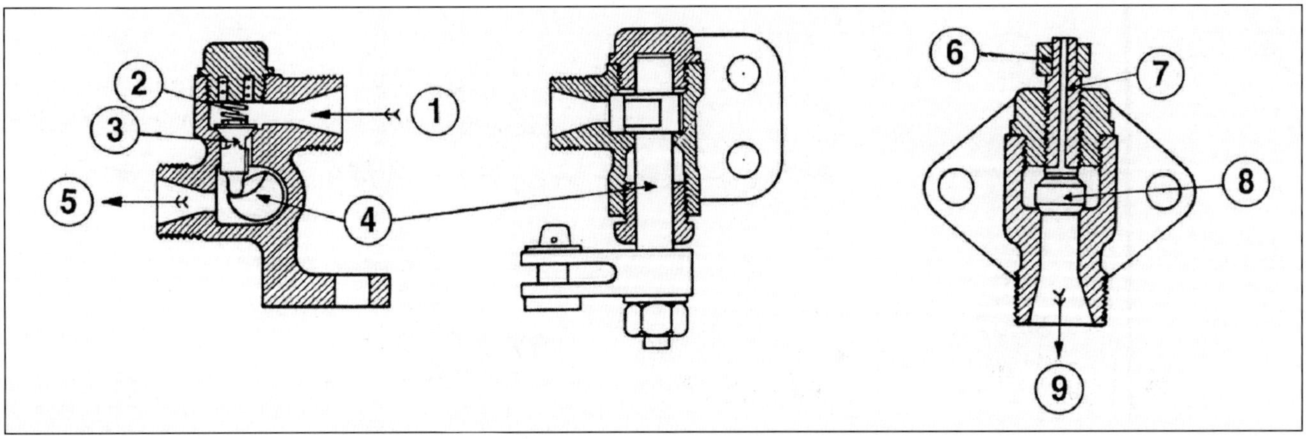

Side and top views of the steam control cock for the atomiser, and the boiler stop valve with its 'tell-tale' drilled spindle that blows when the valve is not wide open: 1 steam supply; 2 return spring; 3 steam valve; 4 operating cam and spindle coupled to cylinder drain cock rodding; 5 steam to atomiser; 6 square for spanner; 7 shut-off screw; 8 steam valve; 9 steam to control valve

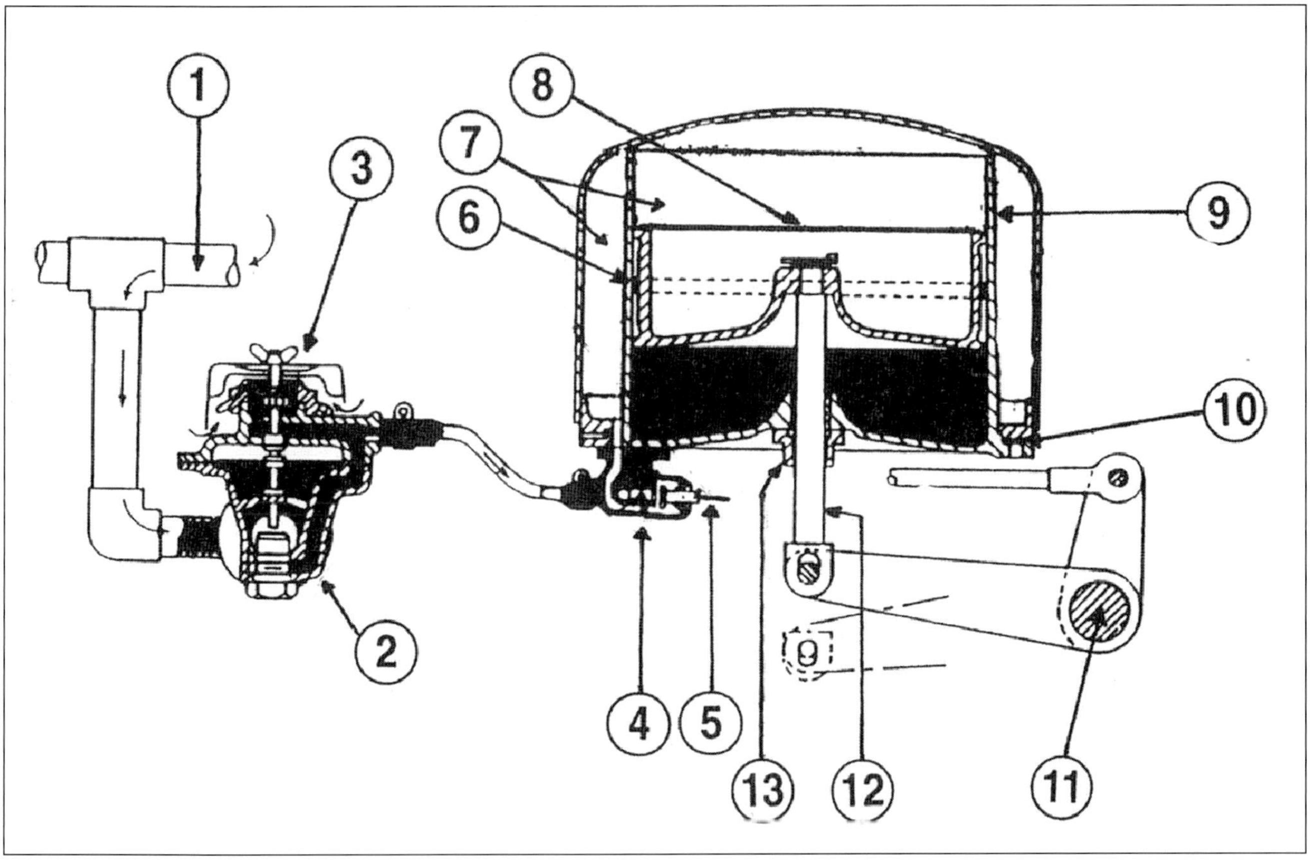

A typical vacuum brake cylinder. To the left is the direct admission valve (2) that allows air into the system to apply the brake. The arrows in the train pipe indicate the direction of the air during an application of the brake, and the brake cylinder is shown in the 'brake on' position: 1 train pipe; 2 direct admission valve; 3 valve open; 4 ball and release valve; 5 connection to release cord; 6 rolling ring; 7 vacuum reservoir; 8 piston; 9 cast iron; 10 sealing ring; 11 brake shaft pivot; 12 piston rod; 13 gland box and gland

of the locomotive, usually on the motion. The oil is stored in boxes or cups, commonly made from brass, which are situated at convenient places on the locomotive. The oil is drained out of the container into the feed pipes by the siphoning effect of worsted wool trimmings. This method of lubrication is usually found where parts reciprocate or slide.

Another type employs a worsted pad seated in a reservoir, which allows the oil to seep slowly to the lubrication point. Another method of control similar to the wool trimming is the restrictor plug, which is made from metal with three flutes and is situated in a feed hole. The restrictor plug type can be found in parts of the motion that rotate, causing the oil to splash around the cup and past the restrictor to a felt pad, through which it feeds slowly to the lubrication point.

TRAIN BRAKES

An Act of Parliament in 1889 decreed that all passenger trains should be fitted with a continuous braking system capable of being applied to all the vehicles of the train. The action should be instantaneous and be able to be applied by the driver or guard at the same time or independently of each other. Should the train become accidentally divided, the brake should automatically be applied, bringing both portions to a stand.

The vacuum brake

Although air brakes were used by the London, Brighton & South Coast Railway, and later by London Transport and the Southern Railway, the most common type of brake used on steam-hauled passenger trains in the British Isles is the continuous automatic vacuum brake, which requires a vacuum to be created to release the brake.

On most steam locomotives the vacuum is created and maintained by an exhausting device known as an ejector. These come in various designs, the most common being the SSJ type, which is described in more detail below.

Former Great Western locomotives use an ejector to create the vacuum in the first instance, after which it is maintained against any leaks in the system by a vacuum pump driven from the crosshead; this was done to save steam.

The automatic vacuum brake relies on atmospheric pressure for its application. When the vacuum is created in the brake system, the piston in each vacuum chamber falls to the bottom by gravity. At this stage there is no air on either side of the piston. To apply the brake, air is admitted to the train pipe, either by the driver or guard operating the brake valve or by a break in the pipe. The air is admitted to the bottom of each vacuum chamber via a device called a direct admission valve. Because there is a vacuum above the piston and atmospheric

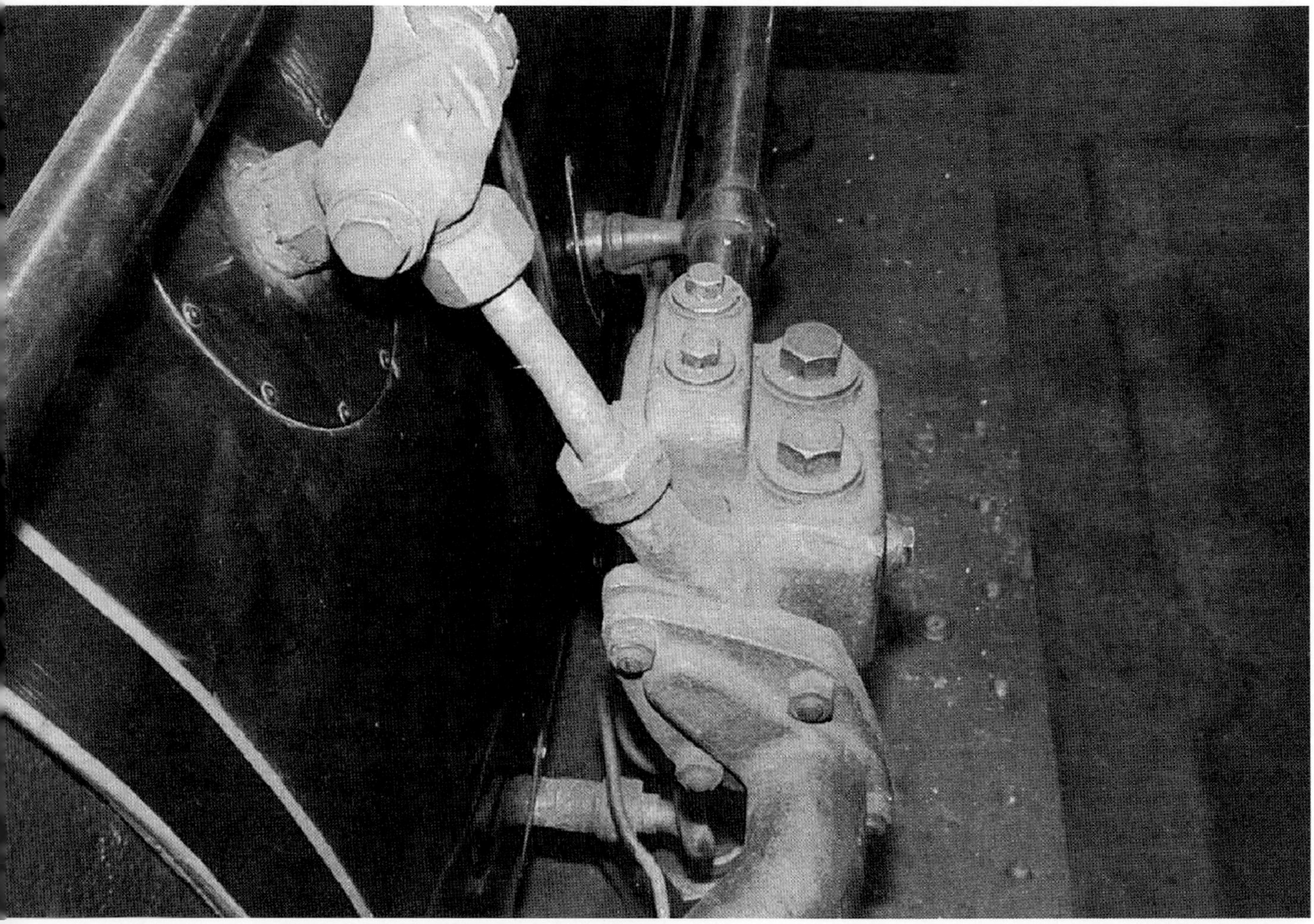

A Midland Railway-pattern ejector fitted to the side of the smokebox. The connection to the large ejector steam supply valve can be seen at the top of the photograph. The cap nuts give access to the steam cones for maintenance. *Author*

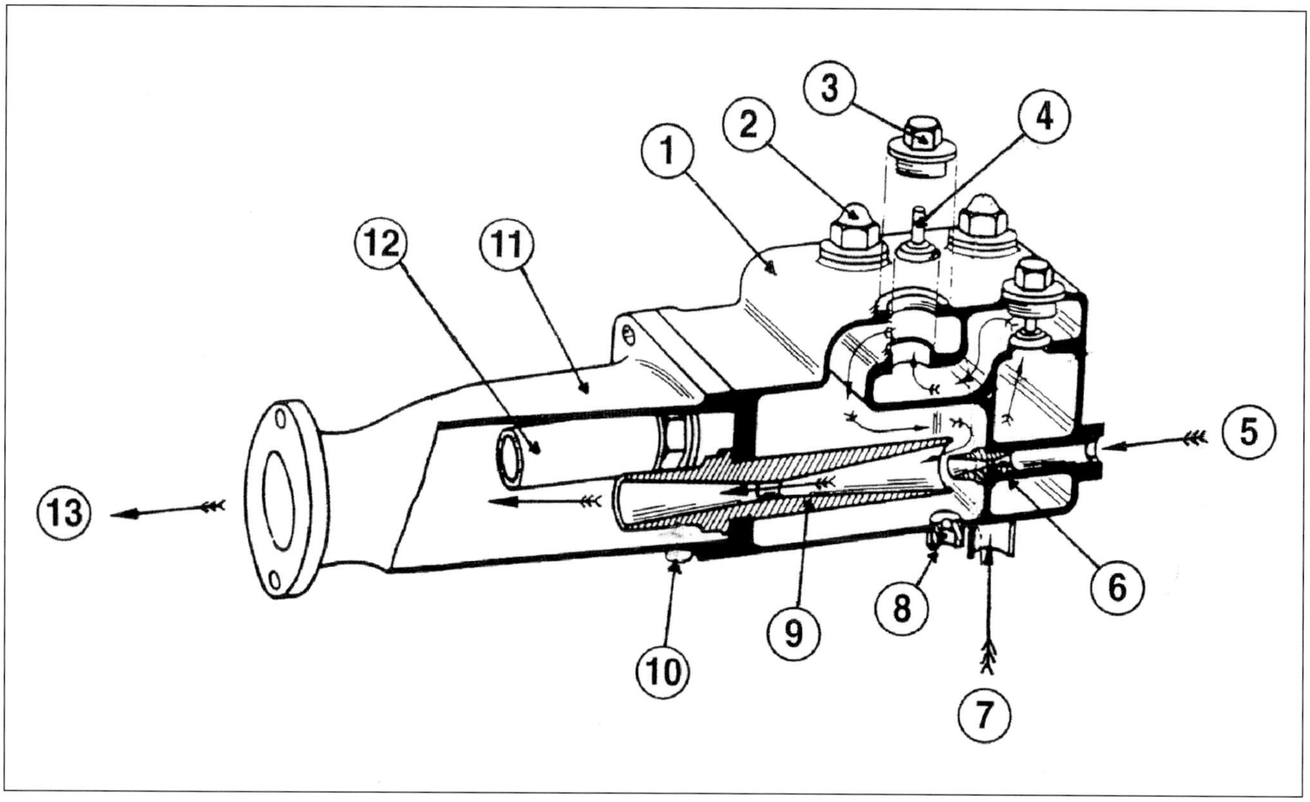

Cross-section of the SSJ ejector. The steam supply from the boiler enters the small ejector steam cone (5) and increases in velocity to the small ejector delivery tube (9), which draws the air from the train pipe and exhausts to the chimney. The same happens in the large ejector: 1 ejector body; 2 large ejector valve cap; 3 small ejector valve cap; 4 non-return small ejector valve; 5 steam supply; 6 small ejector steam cone; 7 from train pipe

An SSJ ejector fitted to a locomotive smokebox, showing on the left the elbow that exhausts the steam/air mixture to the base of the chimney and out to the atmosphere. The large pipe below the right-hand end of the ejector is the connection to the train pipe, and just above it are the two steam connections to the large and small steam cones. The three small pipes drain any condensation from the body of the ejector. *Brian Dobbs*

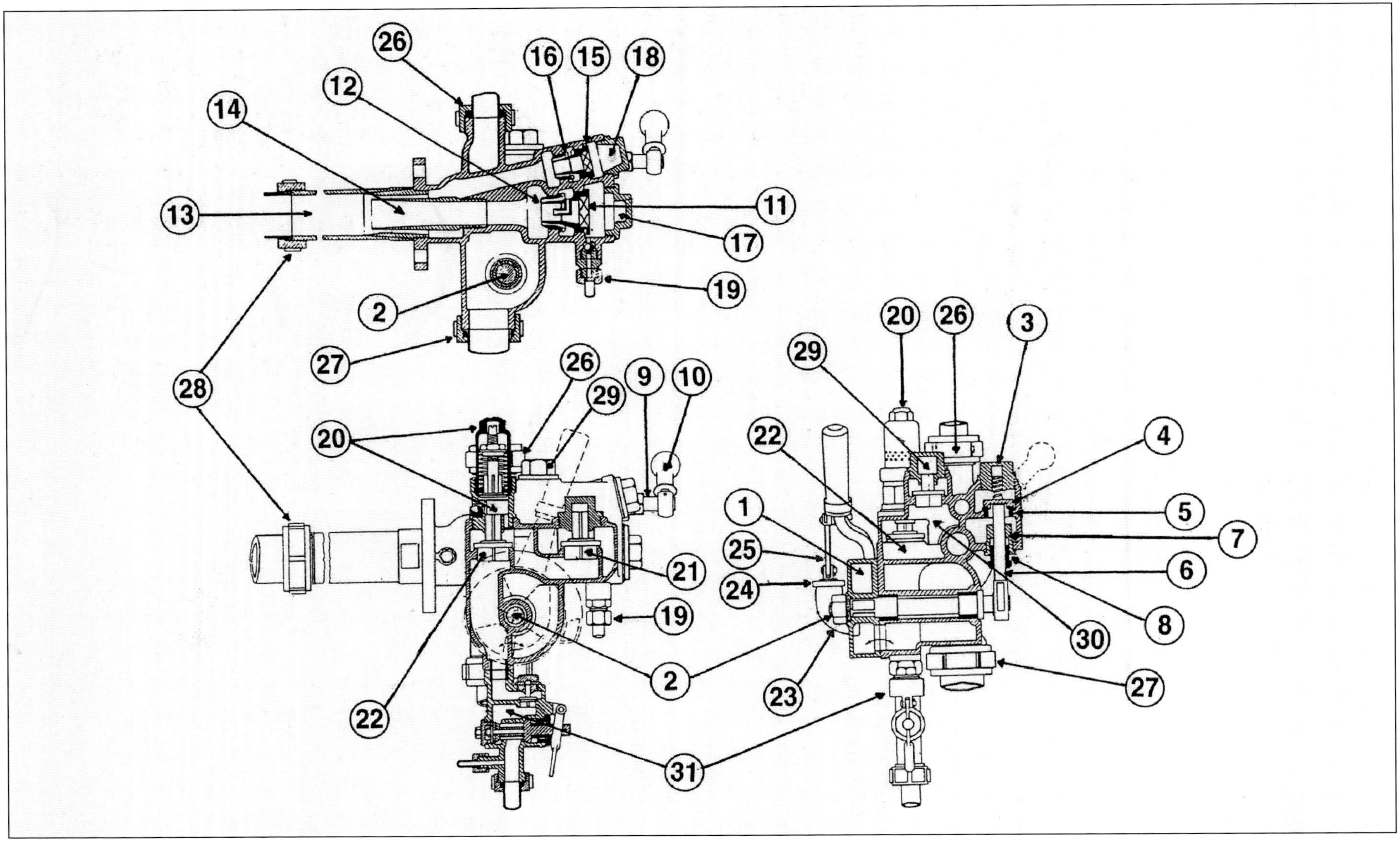

The Dreadnaught brake valve and ejector assembly: 1 air disc with handle; 2 shaft with nut and lever; 3 large ejector steam valve guide nut; 4 large ejector steam valve clock; 5 large ejector steam valve seating; 6 large ejector steam valve spindle; 7 large ejector steam valve packing box; 8 large ejector steam valve gland; 9 small ejector steam valve spindle; 10 small ejector steam valve handle; 11 large cone (inner part); 12 large cone (outer part); 13 ejector exhaust barrel; 14 large ejector exhaust nozzle; 15 small cone (inner part); 16 small cone (outer part); 17 large cap; 18 small cap; 19 drip connection, with nut and ball; 20 vacuum reducing valve; 21 air clack (back); 22 air clack (front); 23 auxiliary application valve; 24 auxiliary application valve clack; 25 auxiliary application valve lever with pins, link and spring; 26 steam pipe union and pipe; 27 train pipe union and ring; 28 exhaust pipe union and ring; 29 small air clack guide; 30 small air clack; 31 release valve

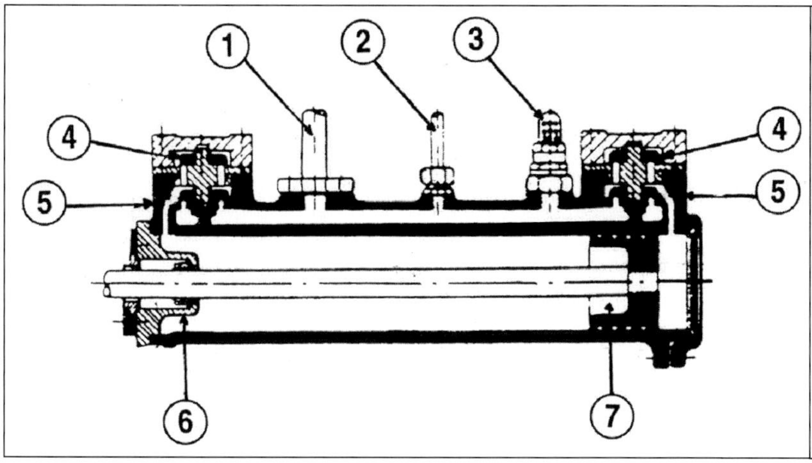

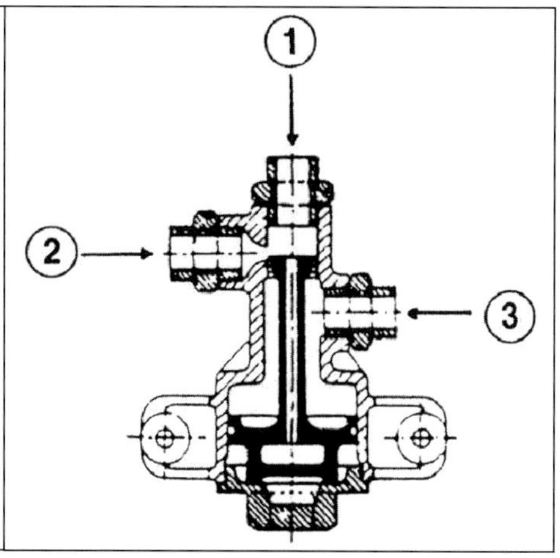

Above left: The Great Western vacuum pump: 1 train pipe connection; 2 lubricator; 3 vacuum adjuster valve; 4 outlet valve; 5 inlet valve; 6 gland; 7 piston

Above right: Vacuum retaining valve: 1 connection to train pipe; 2 connection to pump; 3 connection to reservoir

Below: A Great Western vacuum pump, showing the driving arm from the crosshead. *Peter Herring*

pressure is being admitted beneath it, the piston will rise.

Atmospheric pressure at sea level is about 15 pounds per square inch, which will support a 30-inch column of mercury (vacuum is measured in inches of mercury, 30 inches representing a perfect vacuum). It will be evident from this calculation that 2 inches of vacuum is equal to approximately 1 pound of atmospheric pressure.

The regulation vacuum is between 19 and 21 inches (except on Great Western locomotives, where 25 inches is required).

Ejectors

The **SSJ ejector,** like others, is a steam driven device that removes the air from the train pipe and the vacuum brake system of a locomotive and train. It is fitted to the smokebox or boiler side near the footplate.

All ejectors operate in the same basic manner, but the SSJ is the easiest to describe. It works simply by a jet of steam at boiler pressure passing through a cone to increase its velocity, as in the case of an injector. As the steam passes through the cone it mixes with the air from the train pipe and its velocity draws the air with it into the ejector delivery tube. The air/steam mixture passes through the delivery tube, which has a widening bore, then out into the smokebox via the smokebox elbow. From there it passes to a casting in the base of the chimney and is exhausted to the atmosphere without destroying the partial vacuum in the smokebox.

The monobloc ejector body contains two cones, large and small, which can work independently of each other. The large cone uses a lot of steam and is used to create vacuum quickly in the brake system. As soon as the vacuum is created the large ejector is closed, and the small ejector, which has only a small cone and uses less steam, then maintains the level of vacuum against any leaks in the train pipe connections, etc.

In addition to the cones the ejector body contains two non-return valves that prevent air from entering the train pipe as soon as the ejector steam is shut off. The valves also prevent smokebox ash from being drawn into the ejector. Two further non-return valves at the base of the ejector drain out any condensation that forms in it and prevent a shower of soot from being exhausted up the chimney when the steam is turned on.

The **Dreadnought ejector** has the driver's brake valve and steam supply valves built into a monobloc body. The operation of the Dreadnought is the same as any other design, the vacuum being created by a jet of steam passing through a cone to increase its velocity. As with the SSJ, there are small and large cones forming the small and large ejectors.

As can be seen from the diagram on page 64, on top of the ejector body is the small ejector steam valve (11), while the large ejector steam valve is built into the driver's brake application lever (27). When the lever is placed in the up position it will open the large ejector, but when it is replaced to the running position the large ejector will close and the vacuum will be maintained by the small ejector. The automatic vacuum adjuster valve (21) can be seen on top of the ejector assembly.

The GWR vacuum pump

The Great Western Railway's vacuum pump is driven from a bracket fitted to the crosshead of the locomotive, the back and forth movement operating the piston in the pump. The pump is double-acting, using both sides of the piston.

A Westinghouse air pump as fitted to the smokebox of a Class S160 locomotive. *R. G. Fox*

As can be seen from the accompanying diagram, there are valves at each end of the body. As the piston moves down the cylinder, air is drawn from the train pipe via the inlet valve (1) and a vacuum is created in the chamber on top of the pump body, which is connected to the train pipe. As the direction of the piston reverses, that inlet valve closes and the air is blown out to the atmosphere via the outlet valve (2). On top of the pump body is the vacuum relief valve, which maintains the level of vacuum at 25 inches in the brake system. In the centre is the connection to the lubricator.

A vacuum retaining valve is fitted to the frames of mainly mixed-traffic and goods engines. This was introduced because during long brake applications air admitted to the underside of the brake pistons tends to leak into the vacuum on the top side in addition to the loss of vacuum as the brake pistons rise into that space on first application. The retaining valve transfers the action of the vacuum pump to the reservoir and top sides of the brake cylinders, thus fully maintaining the braking effect. In the same way as the brakes are applied by destroying the vacuum in the train pipe and below the brake pistons, so they are released by restoring the vacuum to the train pipe and the lower sides of the pistons.

LOCOMOTIVE BRAKES

The brake on the locomotive is usually steam-operated; steam at boiler pressure is admitted to a cylinder via the driver's steam brake handle. Most former Great Western locomotives are fitted with a vacuum brake, although some designed mainly for shunting have steam brakes. The vacuum for the Great Western system is created in the first instance by an ejector, and, after a speed of 15mph has been reached, is maintained by the crosshead vacuum pump.

Steam brakes

When the steam brake lever is operated by the driver on the footplate, steam is admitted to a brake cylinder, forcing the piston along the stroke. The piston rod is connected to a bell crank, sometimes called a 'two-to-one lever' (because the arm connected to the piston is twice the length of that connected to the brake stretcher), which pivots in a bracket supported under the brake cylinder.

When the steam brake is operated, pressure is applied to the longer arm of the bell crank, which pulls the brake linkage and forces the brake blocks against the wheels. Steam brake cylinders must be warmed through before attempting to move a steam-braked loco.

Vacuum brakes

A locomotive vacuum brake operates in a similar method to that on the train, except that the cylinder is larger. Under no circumstances should a locomotive fitted only with a vacuum brake be moved without first creating the vacuum, otherwise the locomotive will have only the handbrake operative.

Air brakes

The air brake works in a similar manner to the steam brake. Air is created by a compressor fitted to the smokebox or main frame of the locomotive. Steam is admitted at boiler pressure to the cylinder, causing a piston to rise and fall; a tail rod on top of the steam piston controls a slide valve that admits the steam. Connected to the same piston rod is a piston working in a separate cylinder at the bottom of the casting. As the piston oscillates by the action of the steam in the top cylinder, air is compressed in the lower cylinder. The compressed air is stored in an air receiver or reservoir usually situated between the frames of the locomotive. Under no circumstances must a locomotive fitted with an air brake be moved without first starting the compressor and creating the required amount of air pressure in the reservoir, otherwise (again) the locomotive will have no brake except the handbrake.

Conclusion

The foregoing is just a brief description of the steam locomotive and some of the important aspects of its construction and operation. There are many other specific locomotive designs and developments that have had to be omitted for reasons of space - the Franco-Crosti boiler, for example. However, most such locomotives are no longer in existence or in use in preservation.

We will now proceed to describe the duties of footplate staff from cleaner to fireman, then to the ultimate peak as the driver of a steam locomotive.

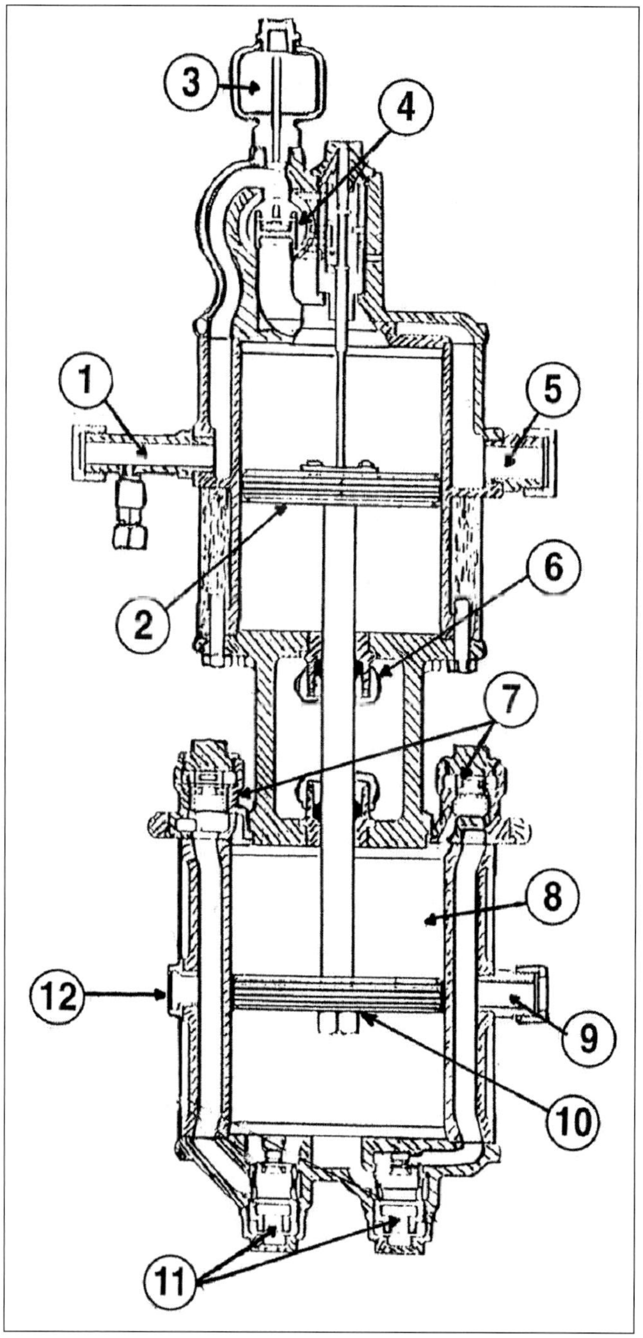

The Westinghouse air pump: 1 steam inlet; 2 steam piston; 3 lubricator; 4 steam slide valve; 5 steam outlet; 6 gland; 7 top air valves; 8 air cylinder; 9 air discharge; 10 air piston; 11 bottom air valves; 12 air inlet

2
Starting out: locomotive cleaning

Most locomotive drivers started their career on the footplate by cleaning locomotives, and this, as now, provided a very useful environment for learning the basics of footplate life.

First, however, as we enter the depot it is worth taking time to consider some important aspects of safety.

SAFETY FIRST

The need for an awareness of personal safety on the part of the staff around the depot of any operational railway cannot be over-emphasised. Every effort must be made by all the working members and full-time staff to maintain a safe working environment. Walkways must always be kept clear of obstructions and free of oil that could cause a fall. Tools should be kept in a safe place and in a serviceable condition; hammers with loose heads can be dangerous; pieces of metal can fly off chisels with mushroomed heads; and worn electrical extension leads can cause electric shocks.

Correct dress

Wearing the right clothing is a most important first step and can help with the safety aspect. Steel-toe-cap boots will protect the feet from heavy falling objects, while good quality industrial glove can protect a cleaner's hands from cuts received from sharp objects and burns from hot surfaces. Rubber gloves give the best protection from skin diseases such as dermatitis, caused by hands coming into contact with oil, and it is advised that these are worn when cleaning the main frames and motion of a locomotive where heavy deposits of grime can be found. Leather industrial gloves give the best protection from burns while working about a locomotive in steam.

The best and most comfortable type of protective clothing is the boiler suit or one-piece cotton overall, which gives the wearer complete protection from the oil and grime that is found around the steam locomotive. Sparks will make holes in nylon overalls, which are not recommended simply because they will not last.

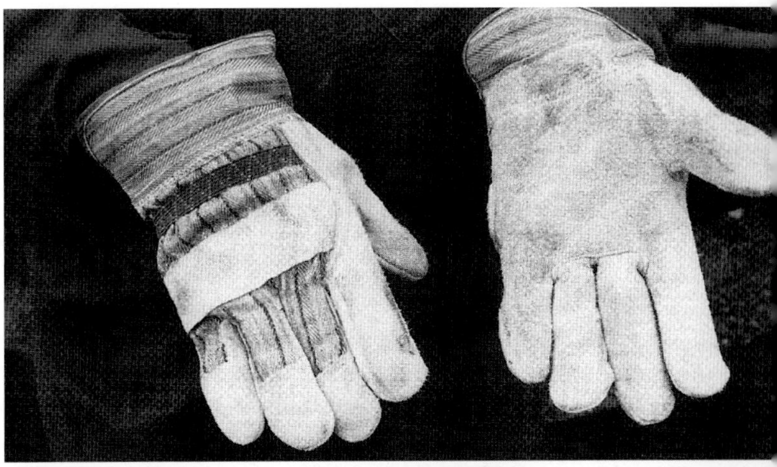

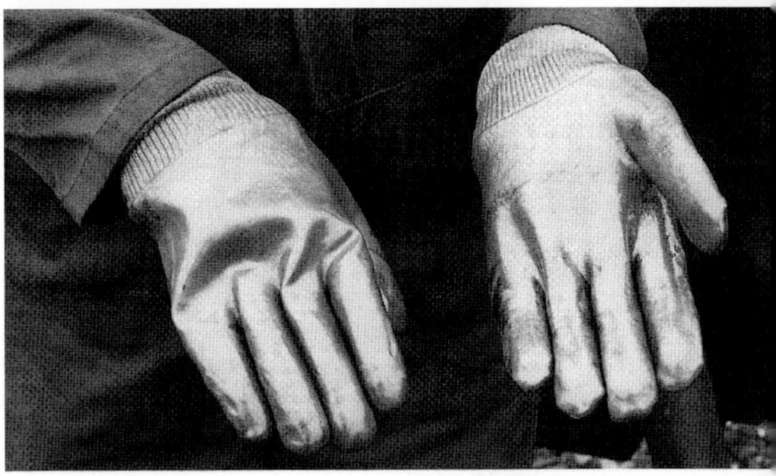

Top: Leather-palmed industrial gloves are best used when working or cleaning near hot surfaces such as pipework on the faceplate or fire cleaning using fire irons. *R. G. Fox*

Above: Rubber gloves are best for cleaning the motion or frame when using cleaning spirit such as degreaser or paraffin. *R. G. Fox*

Reporting to the person in charge

On all operational railways there is someone in charge of the locomotives and the movements that take place about the depot. There should be a book in which to record your arrival at the depot and the purpose of your visit; this record has to be kept in case of an emergency requiring evacuation of the building. It may also be a requirement of the site owner's insurance policy.

It is therefore always necessary to report to the person in charge upon arrival. He will also allocate the locomotive to be cleaned or to have work done on it - after all, there is no point in cleaning a locomotive that the fitters are going to take to pieces or the boiler of which is due to be washed out.

Walking about the depot and along the railway

When walking near the railway your safety depends on your being constantly alert - always use your ears and eyes. In the depot area it is necessary to walk in the 'six foot' between the running lines - never between the rails. A clean orange high-visibility vest must be worn, although this is not necessary when on the footplate of a moving locomotive. There is often the temptation to walk between the buffers of stationary vehicles instead of going round them, which is safer. You should never walk between buffers if they are less than 10 feet apart; if for any reason one of the vehicles is not secured by brakes and is to be shunted, you could become trapped between them, and they come together very suddenly!

When you walk out from the rear of a stationary vehicle, always try to be about 10 feet or more from the end, as this will give a clearer view of any approaching locomotives or vehicles. This action also gives the crew of the locomotive a chance to observe anyone on the line.

Out on the line, always walk in the cess by the side of the running line, never between the rails, on the sleeper ends or

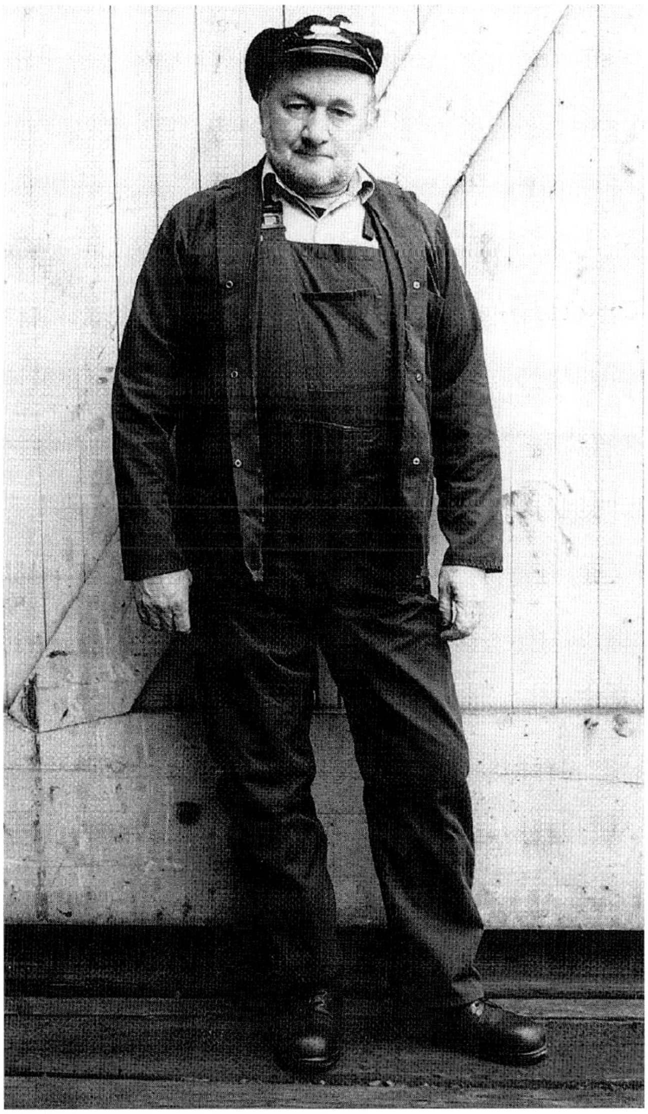

Above left: The typical footplate uniform of 'grease-top' cap, to assist with weather protection, bib-and-brace overalls with a removable slop jacket in case of hot conditions on the footplate, and steel-toe-cap boots to protect the feet from falling objects. *R. G. Fox*

Above right: The one-piece boiler suit is best for protection from oil and dirt when preparing the locomotive; again, steel-toe-cap boots are worn. *R. G. Fox*

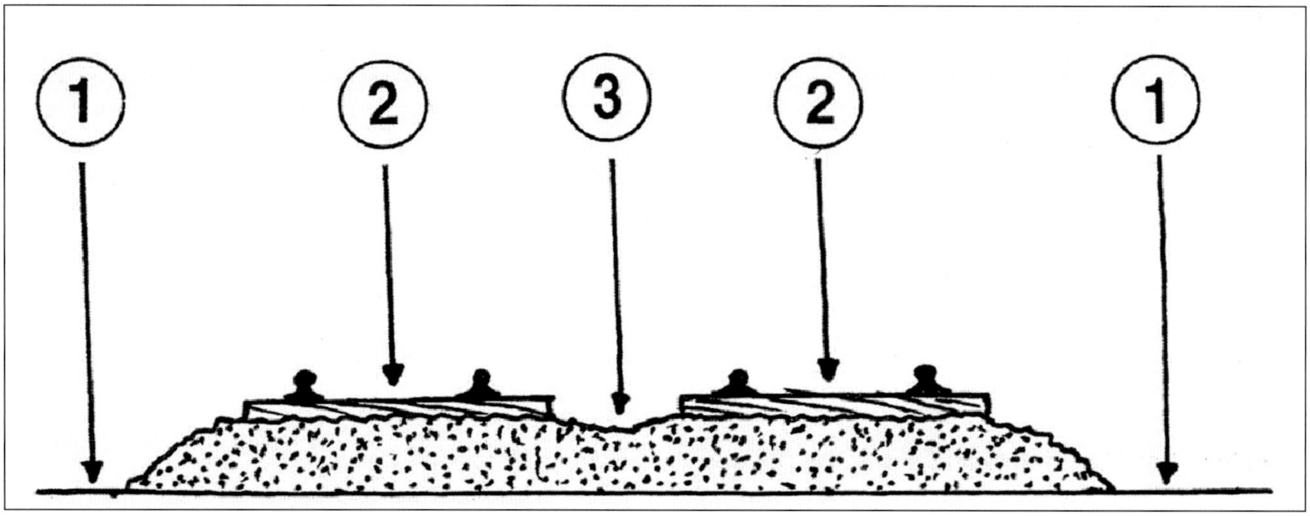

Recognising the various areas of the track formation: 1 the cess; 2 the 'four foot'; 3 the 'six foot'

on the rail head. Keep clear of point rodding, signal wires and other obstructions. Always try to face oncoming trains where possible; when a train appears, stand clear of the track and raise an arm to acknowledge to the crew that you are aware of their existence. Should you be caught in the six foot with trains approaching from both directions, lie down and keep your legs and arms close to your body.

When crossing over the railway always look in both directions to make sure that no trains are approaching, then step *over* the rail, not on top of it, as it could be slippery. Walking through tunnels can be dangerous and they must not be entered without notifying other personnel or seeking advice.

Hand lamps

If you are walking by the railway in the dark or through a tunnel it is advisable to carry a hand lamp, which should be capable of showing red, green, yellow or white aspects. When walking the light must be kept pointing down to the ground, which will indicate your presence to a locomotive crew. Unless the lamp is being used during a shunting operation to signal to other railway personnel, the white light only should be visible; the careless use of any other colour could be mistaken by a locomotive crew and cause the train to start or stop without due cause.

Any colour of light waved violently must be regarded as a danger signal.

Inspection pits

At various points around the locomotive shed and yard are inspection pits used to gain access to the underside of locomotives. Due to their design to accommodate the length of a locomotive they are quite long and take time to walk round, so there is often the understandable temptation to jump across. However, this must not be attempted; a slip could easily result in injuries to the ribs, chest, legs and arms from coming into contact with the rail on the opposite side. Moveable bridge sections should be available to be placed over the inspection pit where convenient and these must be used at all times.

Ear protectors and personal stereos

When working in the depot you may be subjected to excessive noise, so it is often advisable to wear ear protectors. However, when walking about the depot away from the noise these should be removed so that the sound of any approaching train can be heard. It is not advisable to wear personal stereos when walking or working in the depot or near the running line as the warnings made by a train crew might not be heard.

Correct lifting

While working about the railway it is almost certain that items will have to be lifted, and it will be necessary to do this in

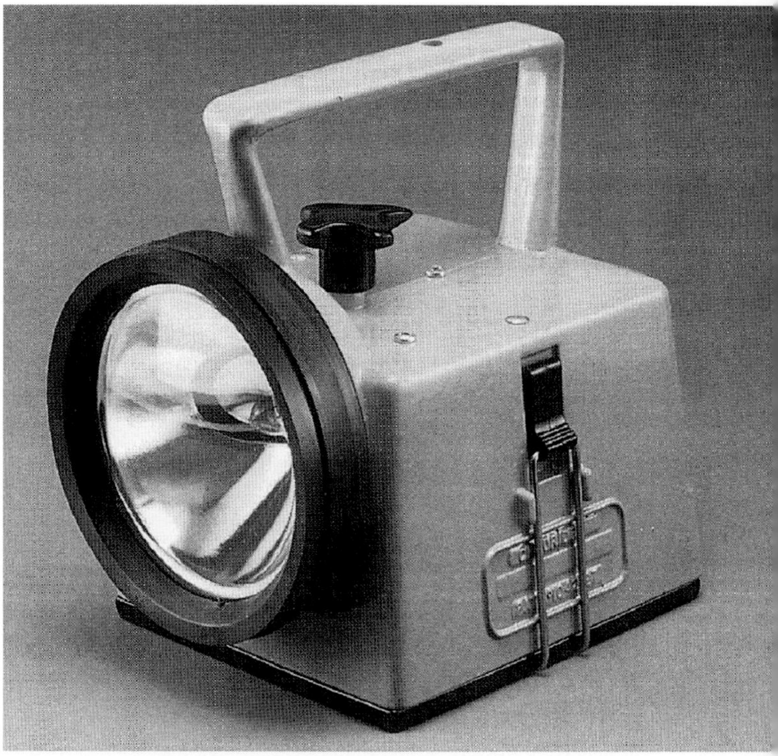

The standard footplate hand lamp. These battery-powered lamps are a very useful part of footplate equipment, and are capable of showing red, green, yellow or white aspects. *Beesley Products*

A moveable board in place over an inspection pit. *Author*

the proper manner. Incorrect lifting can cause slipped discs, lumbago, hernias and injury to the back that can last for years. There are five key points to remember no matter what is being lifted:

- The chin must be tucked in
- The arms *and* back must be kept straight
- The feet must be positioned correctly
- A firm hold must be taken of the object to be lifted
- The thigh muscles should be used to effect the lift.

STARTING OUT

A learning environment

A clean locomotive is not only a pleasure to look at but is more efficient and safer to work with. Time spent by new recruits as cleaners gives them time to learn about the locomotive in general and observe how it is constructed, as well as providing the opportunity to observe the crew going about their duties. The cleaner can talk to the fireman or driver, when they are not otherwise occupied in their work, to ask questions about the engine and gather knowledge about railway operations. As time goes by the cleaner may be able, under the guidance of an experienced fireman, to get some practical experience of footplate duties and the discipline required when out on the line. Riding on the footplate, the trainee will soon appreciate the reason for a clean working area. For example, if coal is left lying around the floor and is stepped on it can cause a twisted ankle.

It will soon be learned that defects on a locomotive or tender, such as a broken leaf in a spring, cannot be observed if there is dirt and grime present, but they could lead to a derailment. A missing cotter-pin may go unnoticed under a layer of grime, causing a vital nut or collar to become loose, again with possibly serious consequences. A trace of rust bleeding from a rivet or bolt head indicates that it is loose or broken, and could cause a part to fall from the engine.

If oil is spilled on the running plate and not cleared away, it could cause someone to slip and fall from the locomotive. Ash from the smokebox is another problem if it is not swept off the running plate, as it may find its way into the driver's or fireman's eyes, causing him to miss observing a signal. It may also blow into the motion of the locomotive, where it will form a very efficient but unwanted grinding paste causing rapid wear to the working parts of the engine.

Safe cleaning

As protection against skin disease it is advisable to apply a good quality barrier cream to the hands before work commences. The locomotive must be fitted with 'NOT TO BE MOVED' boards protruding outwards from the lamp brackets of both engine and tender, or the cab sides. These are to inform personnel that there is someone working about the locomotive. They must be taken from the stores and not from another locomotive! The responsibility for removing the boards always rests with the person or persons working on the locomotive when the work is finished for the day - providing that the locomotive has been left in a safe condition to be moved.

It is advisable to chock the locomotive and to ascertain that the handbrake has been firmly applied. Always keep running plates and steps clear of objects; for example, an oil can might not be seen by someone climbing down from the locomotive and could easily cause a fall. Oil must be cleaned from the locomotive hand rails and from the footplate, otherwise it could cause someone to lose their grip on the rail and slip from the locomotive.

When the locomotive is in steam it must always be borne in mind that, even if you are wearing gloves, the hot surfaces can cause a nasty burn requiring hospital attention, if, for example, a steam pipe is gripped accidentally. The running plate along the side of the boiler is very narrow and usually abounds with hot surfaces such as pipes, lubricators, brackets, etc. Pipes that connect with injectors and vacuum ejectors quite often

A 'NOT TO BE MOVED' board on the lamp bracket of a locomotive being worked on in the depot yard. *Author*

A moveable chock placed behind the wheel of a locomotive. *Author*

run alongside the boiler and if held to assist one's passage along the running plates can cause burns.

While working or walking on the running plate it is advisable to keep one hand on the hand rail at all times and work or carry objects with the other.

Safety on the footplate

When a locomotive is moving, a number of potential dangers exist on the footplate, all of which can be made safe by the careful conduct of the crew.

The most common injury is to the feet, but, as stated previously, safety boots can eliminate most of the danger. However, care needs to be observed at all times when on the footplate. On a tender locomotive there is a hinged steel plate, known as the lip or fall plate, between the engine and tender. The movement of the locomotive, particularly when negotiating a curve, causes this to slide along the tender floor, and it is possible to get the heel of a shoe caught between the plate and the hand brake pillar, resulting in injury to the ankle.

Another possible cause of an accident are the footplate doors on a tender locomotive. It will be noticed that when the engine is in motion and the doors are closed, the space beneath them alters when the locomotive passes over rail joints. It is therefore important never to place a foot in the gap between the footplate and the bottom of the door.

No one should ride on the steps of a moving locomotive for two main reasons: should the person slip and fall he could of course end up under the wheels of the moving locomotive, and if the locomotive passes an object with limited clearance the person could be crushed.

Further causes of accidents are scalds and burns. These can be caused by a gauge glass blowing out and showering the footplate with steam and water; the shattered glass is retained inside the gauge frame protector, but the steam escapes on to the footplate. If this should happen, stand back out of danger until experienced staff can deal with the problem. In the event of no experienced member of staff being present, it will be necessary to report the defect to the person in charge of the locomotives as soon as possible.

LOCOMOTIVE CLEANING

Main frames and wheels

The main frames of the locomotive are the most difficult part to clean as they collect all the dirt and grime from the track, together with ash from the ashpan. Large areas of the frames cannot be reached easily as they are concealed behind the wheels and motion. A steam cleaner can be used, but it condenses on the frames and after cleaning it will be necessary to drain the water out of the oil cups and axleboxes. It must be remembered that the oil will float on top of the water, so at first glance there may appear to be no water in the cup. The water can be removed by sucking it out from the bottom of the cup using a syringe.

Another method of cleaning the frames is with a mixture of oil and paraffin or red diesel oil. Use an old but clean paintbrush to apply a generous amount of the fluid to loosen the grime from the frames, then use a good handful of cotton waste to remove the residue. A shine can be brought up on the paintwork of the frames and any of the oil left behind removed.

The wheels can be cleaned by the same means.

Motion

Rust on motion parts (valve gear, coupling rods, etc.) can be cleaned off with a mixture of oil and paraffin in conjunction with a medium-grade emery cloth. Place the oil and paraffin mix on the part of the motion to be cleaned, then, using the emery cloth, spread the mixture along the motion. Then, by using a sliding movement with pressure applied, polish an appropriately sized area of the motion. An abrasive black residue from the emery cloth will be left behind, and this must be washed off with clean paraffin.

If the part worked on is still not clean enough, it may be necessary to repeat the action until the required level of polish has been achieved. Attention must be given to prevent the mixture of oil and paraffin from entering the pivoting part of the rod. If this should happen it must be washed out with clean paraffin or red diesel oil and then re-lubricated to ensure that no damage has been caused. Heavily tarnished motion can be cleaned with a sanding disc or flap wheel on a low speed power drill.

The polish on the motion parts can be protected from rust after cleaning by the use of a mixture of cooking oil and paraffin. As a rough guide, the mixture will consist of about 30 per cent paraffin and 70 per cent cooking oil. When applied to the motion the paraffin will soon evaporate, leaving behind a thin film of oil that will protect the shine. After cleaning, this mixture can also be used to protect the main frames and wheels.

Paintwork

The paintwork around the boiler, cab sides and tender can be polished with a good-quality car polish, following the manufacturer's instructions.

Before any attempt is made to polish the paintwork, it will be necessary to remove particles of grit from the surface with a soft brush; make sure that the dirt is not brushed into any bearing surfaces. Any areas of oil on paintwork can be cleaned off with turpentine.

Never dry-polish dirty paintwork, or scratches will be left on the paint surface; it must be washed first, using a wash-and-wax shampoo or a household detergent such as washing up liquid in hot water. Starting at the highest point of the boiler or tender, soak the surface with plenty of soapy water. If the locomotive is warm, wash only a small area at a time to prevent the streaking that will occur if it dries too quickly.

Another method of cleaning the paintwork is, again, a mixture of engine oil and paraffin or red diesel oil. This was used in the final days of steam by British Railways and, although it afforded some sort of a polish and protection, it did tend to attract any dust and dirt given off by the locomotive, so never looked clean for very long.

When using a ladder to clean the top of the boiler it must be secured against the hand rail by hooks that clip over it to prevent it from sliding away at the bottom, causing the cleaner to fall.

Brass and copper

Brass and copper always look smart when cleaned and polished. They can be cleaned with steel wool and any good-quality metal polish. If the brasswork on the gauge frame is tarnished it can be cleaned with emery cloth, a flap wheel or sanding disc.

When the pipework around the footplate has become tarnished, perhaps during a period of standing after a heavy repair, it can be cleaned with a fine-grade emery tape and thin oil. It is important to use the oil in conjunction with the tape to lessen the abrasive effect of the emery, which would damage the wall of the tube. Pour an amount of the oil on to the pipe to be cleaned, then take a length of tape about a yard long and wrap it around the pipe. Cover the tape with the oil, then, holding the ends, pull it backwards and forwards so that the tape grips the pipe. Soon the pipe will be covered with a black film of oil mixed with dirt, and it will not be long before a good shine is achieved. If the copper is still not clean enough the process can be repeated. This method can be finished by using a paste made up from powdered pumice stone and oil.

An alternative is a mixture of domestic powdered pan cleaner and oil applied in generous quantities with a long piece of rag, using a similar action to the emery tape and oil.

When polishing the pipework of a locomotive in steam, leather gloves must be worn to protect the hands from the burns that will inevitably come from contact with the hot pipework.

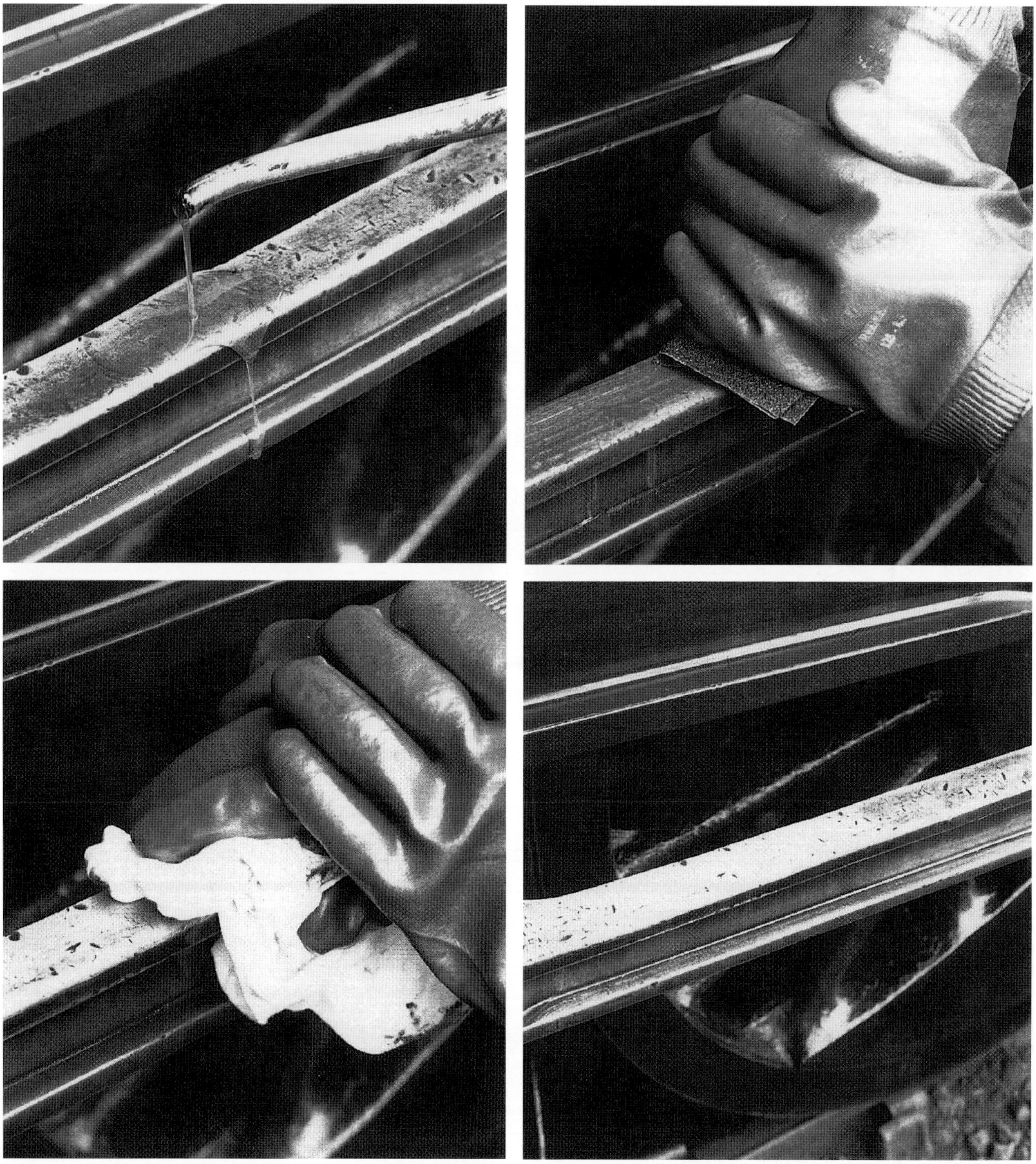

Cleaning a steel motion part. First the oil and paraffin mix is poured on to the surface, then spread with an emery cloth, which is used to clean the part. This action will leave a black film of dirt and abrasive from the cloth, which must be kept away from any bearing surfaces. The job is completed by washing off with paraffin and polishing up with a clean rag. A coating of clean engine oil will protect the surface from future rust, but this will soon attract dirt. *R. G. Fox*

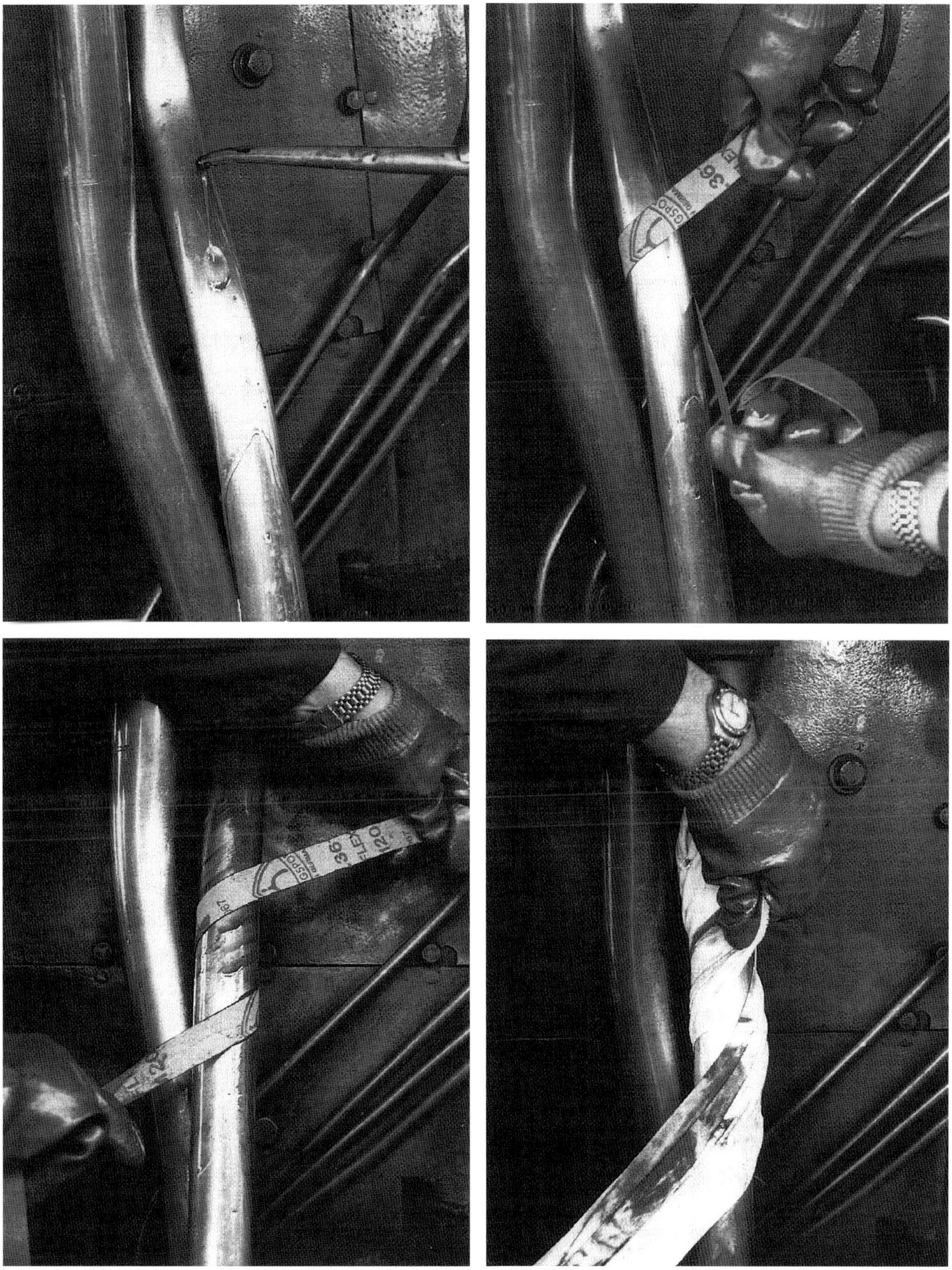

Oil is used on a copper pipe to protect it from excessive damage from the abrasive action of the emery tape. The tape is wrapped round the pipe then pulled backwards and forwards to remove any tarnishing from the surface of the copper. The pipe is then cleaned with a clean rag in the same manner. *R. G. Fox*

3
The fireman's duties: preparation

The duties of a fireman are to supply steam for the locomotive, to assist the driver and to keep the footplate clean. When not engaged in his actual firing work he must also keep a good lookout and observe signals.

While working on the locomotive the fireman has a good opportunity, as he did as a cleaner, to observe the driver going about his duties and thereby gain further knowledge of locomotive driving. At all times the fireman will be under the direct supervision of the driver. If the locomotive is to perform efficiently it is essential that the driver and fireman work as a team, whether preparing the engine or out on the line.

While the driver makes his first visual inspection of the locomotive the fireman must collect the lubricating oil, and if the locomotive is in steam, place the oil on the warming plate ready for the driver; this will enable it to flow more easily.

CHECKING THE BOILER WATER LEVEL

Gauge frames

When a locomotive is being prepared, the first thing to check on entering the cab is that there is sufficient water in the boiler. This is done by operating the water gauge frames on the boiler faceplate.

First it must be ascertained that the two gauge frame shut-off cocks leading from the boiler are open. The cock at the top should be in the 'up' position with the one at the bottom pointing downwards. On more modern locomotives these two levers are linked together by a common rod so that both shut-off cocks can be opened or closed at the same time.

Pull the test cock at the bottom of the frame fully outward through 90 degrees, and the water level will drain from the

The two gauge frames on the boiler faceplate of a BR Standard locomotive. It can be seen that the shut-off cocks are joined together by a common rod, and that the lever is in the fully open, 'up' position. The single lever that closes the frame test cock is at the bottom. *Brian Dobbs*

glass until none can be seen. Close the test cock, returning the handle to the vertical position, and the level of the water in the boiler will be repeated in the gauge glass.

Most locomotives have two gauge frames fitted to the boiler faceplate; if, after both have been tested, two different levels are shown, the lowest one must be taken as being accurate, and the defect reported to the driver.

Former Great Western locomotives are fitted with only one gauge frame, but the method of cleaning and reading the water level is just the same. However, the GWR design has two test cocks fitted to the body of the frame. Should the content of the boiler be suspect when testing this type of gauge, the water level can be verified by operating the top water test cock. If there is water in the boiler it will be seen to flow through the cock when it is opened. However, if no water flows when this cock is opened the bottom one is then used, and if water passes through this in a constant flow then the water level is sufficient to light the fire. There must be about half a glass of water showing in the gauge before any attempt is made to light the fire.

If the gauge glass or protector is dirty it makes it difficult to read the internal glass tube. The protector can be cleaned by the following method: close the two shut-off cocks on the boiler faceplate, then open the drain cock to let the water out of the gauge glass, and leave it in the open position. The gauge glass protector can now be removed by turning it round on the gauge frame so that the retaining clip can be taken out. These vary in type from a single pin that runs the full height of the gauge frame protector, to a clip that passes across the back of the protector. The BR Standard type uses the GWR pattern, which has four lugs or hooks on the back that retain the protector. The glass can now be cleaned with a clean rag while any hard deposits from the boiler salts on the glass can be removed with a piece of steel wool.

Turning to the gauge glass protector itself, it will be noted that the metal door has diagonal slots with a wire mesh. This is to allow the air to circulate around the glass and prevent condensation building up inside the protector frame. The metal of these slots should be painted white to give the effect of diagonal black and white lines. This is done so that the water level in the glass will appear to reverse the direction of the lines by the refraction of the water. Occasionally these

A Great Western gauge frame, showing the nut at the bottom of the frame that is level with the firebox crown sheet. The two test cocks at the rear are used if the gauge glass should blow. It will be noted that the shut-off that isolates the frame for cleaning and the normal drain cock are the same or similar to the pattern that is fitted to the locomotives of other railway companies. *Brian Dobbs*

At night the gauge glasses are illuminated by a gauge lamp. Due to the heat rape oil rather than paraffin must be used; if that is not available an alternative is a mixture of engine oil and paraffin, although it is not recommended. *Beesley Products*

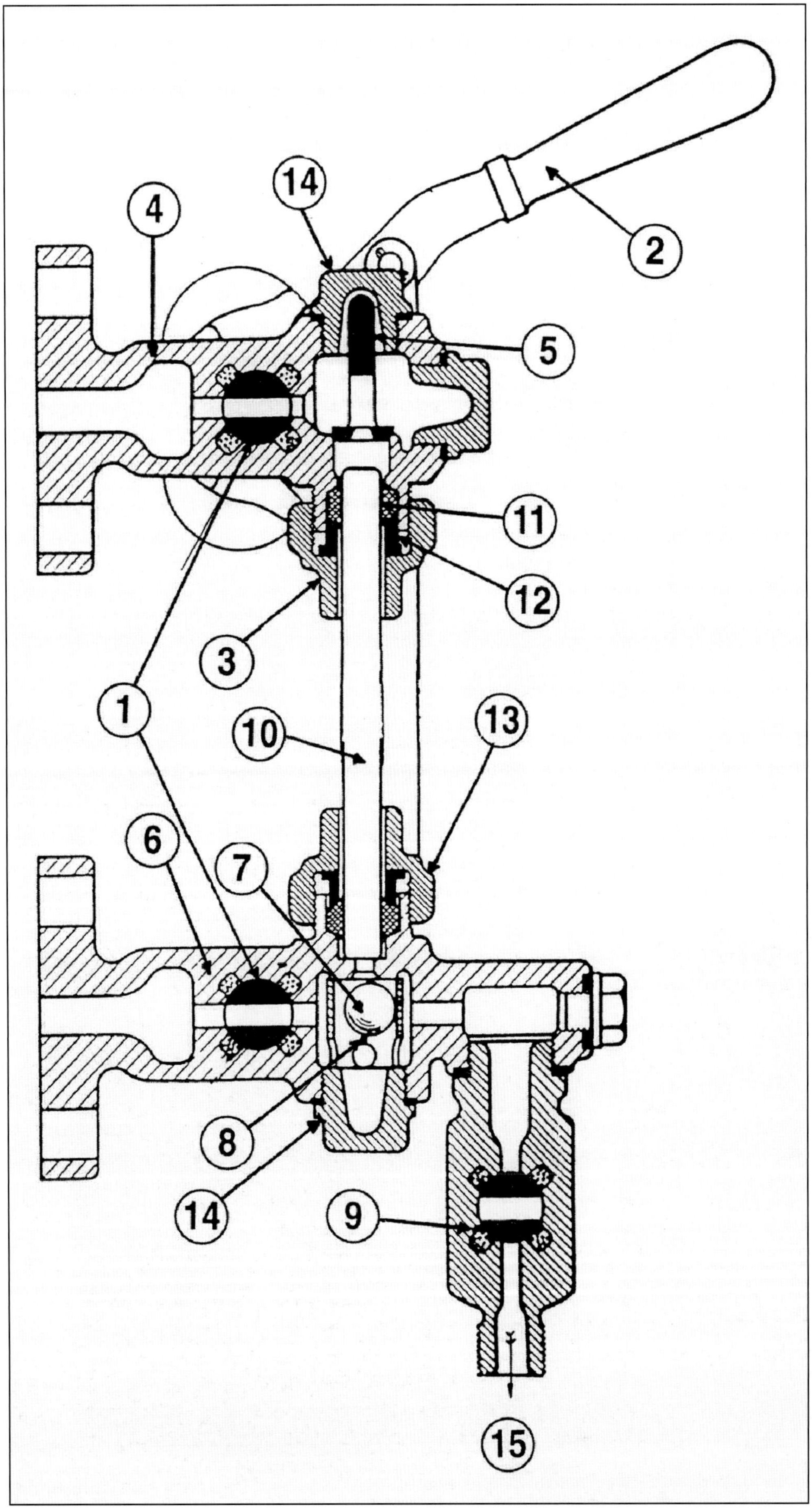

Cross-section of the gauge frame with the protector removed: 1 asbestos packing; 2 shut-off lever; 3 top nut supporting protector; 4 top casting; 5 restrictor plug; 6 bottom casting; 7 ball valve; 8 seating; 9 drain cock; 10 gauge glass; 11 sealing ring; 12 sealing collar; 13 bottom nut that carries the protector; 14 top and bottom cap nuts; 15 to drain air holes are circular, but whatever their shape they fulfil the same purpose.

When the glass and protector have been cleaned they can be replaced on the frame and the retaining clip secured. The frame can then be re opened: with the drain cock still open, the two shut-off cocks are opened slowly so that the gauge glass is warmed before pressure is introduced. The drain cock is then fully closed, and the water level will be seen to rise in the glass.

If no water can be seen in the gauge glass it must be assumed that the boiler is empty and therefore it must be filled before any attempt to light the fire is made. It may be that the boiler is so full from the last time the locomotive was in steam that the water will not come down the glass. This can be verified by closing the shut-off cocks and slowly opening the test cock. If there is any water in the boiler it will be seen to fall in the glass. Close the test cock and slowly open the shut-off and the water will be seen to re-appear in the glass.

It is possible for the gauge glass to blow out while the locomotive is in service; this is caused by wear and damage due to the dramatic changes in temperature. When the glass blows the shattered pieces should be retained by the protector. Inside the gauge frame casting there are two ball valve restrictors that reduce the discharge of steam and water on to the footplate. The blown frame must be closed and the other utilised until the glass has been replaced. In the case of Great Western locomotives with only the one gauge frame, the water level can be tested by using the test cocks on the side of the frame. Gauge frame glasses will need to be changed periodically, such as when the boiler is washed out or inspected.

The procedure for changing a blown glass is shown in step-by-step photographs over the next few pages; refer also to the accompanying exploded diagram. First check the water level in the boiler and refill if necessary. Next close the frame using the shut-off cock (2 in the diagram), leaving the test cock open, and remove the protector frame as described above. Slacken and remove the two collared nuts (3 and 13) that carry the glass and place to one side; inside each will be found a circular rubber sealing ring (11) in which will usually be found the remnants of the gauge glass (10). On the top and bottom of the frame are two cap nuts (14); these are now removed. Check that there is no broken glass left in the castings that could obstruct the water, and replace the bottom nut.

The two sealing nuts (3 and 13) are placed around the replacement glass with the threaded parts outwards. Next fit new rubber sealing rings (11) complete with collars (12 - if fitted) to each end of the gauge glass and slide the nuts and rings to the centre of the glass. If no new seals are available, the pieces of broken glass will have to be removed from the existing ones.

Dismantling and replacing a gauge glass. *Photographs by R. G. Fox. Below left:* Close the gauge frame by pulling down the shut-off cocks. Leave the bottom test cock open. *Below right:* Remove the pin or clip that retains the gauge glass protector.

Lift off and remove the protector from the casting.

Slacken the two collared nuts that carry the glass and leave loose.

Slacken and remove the top cap nut and lift out the restrictor from the top casting of the gauge frame. Take care not to mix up the top and bottom nuts as there may be a difference in the recesses for the restrictor and the ball valve. One way to avoid any confusion is to mark the top nut with chalk, but (if this is not available) put this nut in one of the lockers and leave the bottom one on the warming plate.

Lift the gauge glass complete with the sealing rings, collars and gland nuts clear of the bottom casting, then pull forward and remove from the assembly.

Above: Check that there is no obstruction caused by broken glass or boiler salts restricting the ball valve in the bottom casting.

Above: The components of the gauge frame in order ready for re-assembly. Note that the large collared nuts with the spigot to carry the protector are in the centre of the glass.

Below: Replace the glass assembly complete with the nuts, collars and seals, first into the top casting then down to the bottom seat. Slide the lower nut with the sealing ring to the threaded part of the bottom casting and finger tighten it, then finger-tighten the top nut.

Below: Hold the glass down to the bottom seat and tighten the lower nut with a spanner, followed by the top nut. The nuts must be steamtight but not over-tightened.

Above: Replace the restrictor into the top casting with the seat downwards and the spigot upwards. This locates in the recess in the top cap nut. Replace the top nut and tighten.

Below: With the test cock fully open, gently open the shut-off cocks. This allows the gauge glass to warm up. Keep the test cock open for about 10 seconds, then close it slowly. The water will rise in the glass.

Above: Replace the gauge glass protector and secure with the retaining pin or clip.

Below: Where two gauge glasses are fitted, check and compare the two levels - they should read the same.

With the top cap nut (14) still removed, tilt the glass assembly and place it in the top of the frame. Lower the glass into position against the stop at the bottom of the frame. Slide the nuts (3 and 13) to the threaded parts of the frame and make each finger tight, then tighten with a spanner. The top cap nut can now be replaced, and the frame opened as before. Check that the frame is watertight and steamtight, and compare it with the other gauge to check for a correct level.

Filling the boiler

The filling of the boiler can be achieved only if the locomotive is not m steam and is cold. One of the washout plugs needs to be removed (these can be identified by their square head and taper thread) and the boiler filled by means of a hose pipe.

The plug to be removed needs to be above the level of the firebox crown sheet. Looking at the bottom casting of the gauge frame, the head of a hexagon bolt will be seen at the front, just above the test cock. The firebox crown sheet is directly in line with this. The removal and replacement of a washout plug is not a task for a fireman or driver, but must be done by a fitter who is authorised for safety critical work, as per The Railways (Safety Critical Work) Regulations, 1994.

Larger locomotives can usually be filled via the injector overflow pipe.

CHECKING THE FIREBOX

Due to the different designs of locomotive and their various intended uses, fireboxes come in many different shapes and sizes. Most are long, narrow and deep in design, fitted between the main frames of the locomotive. A small locomotive designed for shunting will have only a small firebox, while a modern engine may have a square and therefore wide firebox mounted over the main frames. Consequently these are more shallow.

Firebox doors

Firebox doors vary in design from the basic sheet steel horizontal sliding type, as fitted to industrial tank engines, to the hollow cast iron sliding version on large passenger

The fire door arrangement of a BR Standard locomotive, showing the handle and levers that control the doors. At the bottom can be seen the hinged flap of the half-moon. plate that controls the admission of secondary air. Just above is the cast iron protector ring. *Brian Dobbs*

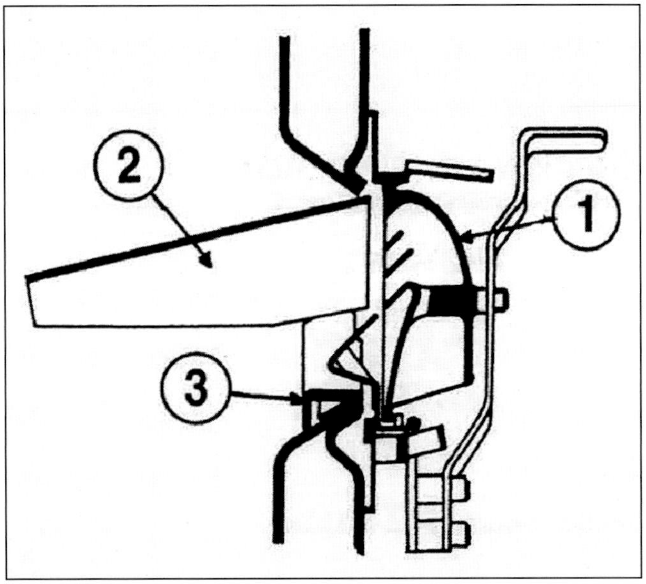

Cross-section of the fire door arrangement showing the air space and the half-moon plate in the up position: 1 fire door; 2 baffle plate; 3 fire door protector ring

locomotives. The doors are mounted in slides at the top and bottom and operated by levers from one handle.

On, for example, an industrial locomotive the firebox door simply keeps the heat in the firebox and provides access to place coals on the fire, but on larger locomotives the doors also regulate the admission of secondary air to control the heat of the fire. To vary the supply of secondary air the doors can be partly or fully opened, while hollow doors provide a controlled supply of air to assist combustion when in the closed position.

Fitted just under the doors on some locomotives, particularly on LMS and BR Standard types, is a steel plate known as a half-moon plate, which pivots to fit inside the doors when they are closed to reduce the supply of secondary air if required. This causes the primary supply of air to be drawn up through the firebed and increases the combustion of the fire.

Former GWR locomotives have a hinged flap plate that works in a similar method. It is flatter and larger than the LMS/BR type, has holes to admit secondary air if required, and is operated by a chain.

Some of the larger Southern Railway locomotives had steam-operated doors that pivoted at the top, controlled by a pedal, so that they could be opened and closed between shovels-full. Other SR doors were operated on a ratchet and opened inwards, with a hinged plate at the bottom that controlled the secondary air.

Larger LNER locomotives had the biggest fire doors, similar in design to those fitted on stationary boilers. The door opened on a hinge and the fire was made up before departure. This large door could not be opened later, and the fire was maintained when the locomotive was working by feeding coals through a hinged flap in the door itself. The flap was controlled by a ratchet and also admitted secondary air into the fire.

Firebox interior

A good guide to the condition of the boiler and firebox is that if there is still enough water in the boiler from the last time the locomotive was steamed to enable the fire to be lit, everything should be in order.

With sufficient water in the boiler, a check must be made of the inside of the firebox. To do this, remove the baffle plate from inside the firebox door and, using a lamp, inspect the condition of the interior. The lap joints of the tubeplate and firebox door plate must be checked for any sign of water leaks. Ideally the firebox should be completely dry, although a damp patch around the joints, giving evidence of water seepage, can be acceptable.

With the baffle plate removed it should also be possible to inspect the condition of the tubeplate above the brick

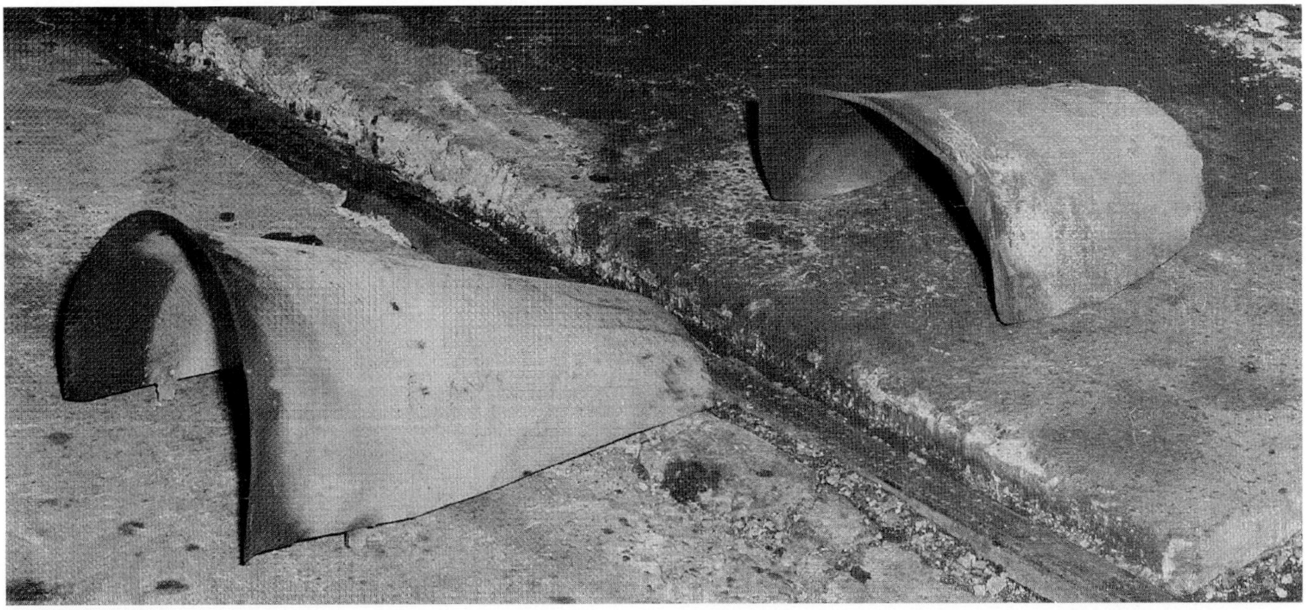

Two baffle plates. The one to the left is in good condition, while the one on the right is wasted, causing poor deflection of air in the firebox. *Brian Dobbs*

arch. There may be a trace of verdigris, which indicates a slight seepage of water from the tube ends when the boiler cooled down after the last time the locomotive was steamed. The fusible plugs must be checked for leaks or drips of water, and if there are any present the driver must be notified or a fitter summoned.

Next to be checked is the brick arch and the baffle plate itself for signs of wasting.

All the clinker must be removed from the firegrate by shovelling it out with the paddle. It may be found easier to climb into the firebox, lift out two or three firebars from the grate and push the clinker out through the space into the ashpan; before doing so, however, it will be necessary to put on a mask to protect the lungs from the dust and goggles to protect the eyes. When the firebars are replaced it is important that they are the correct way round and placed back in the grate in the same position as they were removed. The firebars themselves should be checked for any excessive distortion or burning, as this may cause starvation of primary air to the fire that would also be a possible cause of a bad-steaming boiler.

Boiler tubes

The flue tubes must be checked to ensure that they are clear of small pieces of ash and clinker that could obstruct the flow of gases from the fire. This would also cut down the heating area of the tubes and impair the steaming of the locomotive.

There are two methods of checking the tubes. The first can be carried out from inside the firebox, assuming of course that there is no fire. If the smokebox door is opened, light should be seen through the tubes when looking from the firebox tubeplate. The second method is used when there is a fire in the firebox; if the smokebox door is opened the glow of the fire should be seen through the tubes when looking from that end.

If the tubes are obstructed they can be cleared by the use of a specially designed round wire brush. This is attached to a long flexible steel rod that enables the brush to be pushed right through the tubes to clear any obstruction. This is best performed by pushing the brush from the smokebox end to the firebox.

Attention should also be given to the superheater flue tubes, although these cannot be cleared by the brushing method because of the presence of the superheater elements. However, these do not usually present so much of a problem with blockages due to their larger size.

Rocking grates

Certain locomotives have a drop or rocking grate to ease the task of cleaning the fire.

The early design of rocking grate consisted of a hinged flap that formed part of the grate and which could be opened to allow the clinker to be pushed into the ashpan. Some LNER locomotives had a section at the front of the grate that opened for the removal of clinker, while later LMS and BR locomotives were fitted with a grate that was split into front and rear sections that enabled the complete fire to be removed without the use of fire irons.

Such grates are fitted with two-way locking clamps on the footplate. With the clamp in the first position the grate is partly opened so that the clinker can be broken up while the locomotive is in motion. When the clamp is in the second position the grate can be fully opened to remove the clinker into the ashpan.

The ashpan

When it is ascertained that the firebox is in a satisfactory condition, attention must then be given to the underside of the locomotive, and the ashpan. This should be clear of

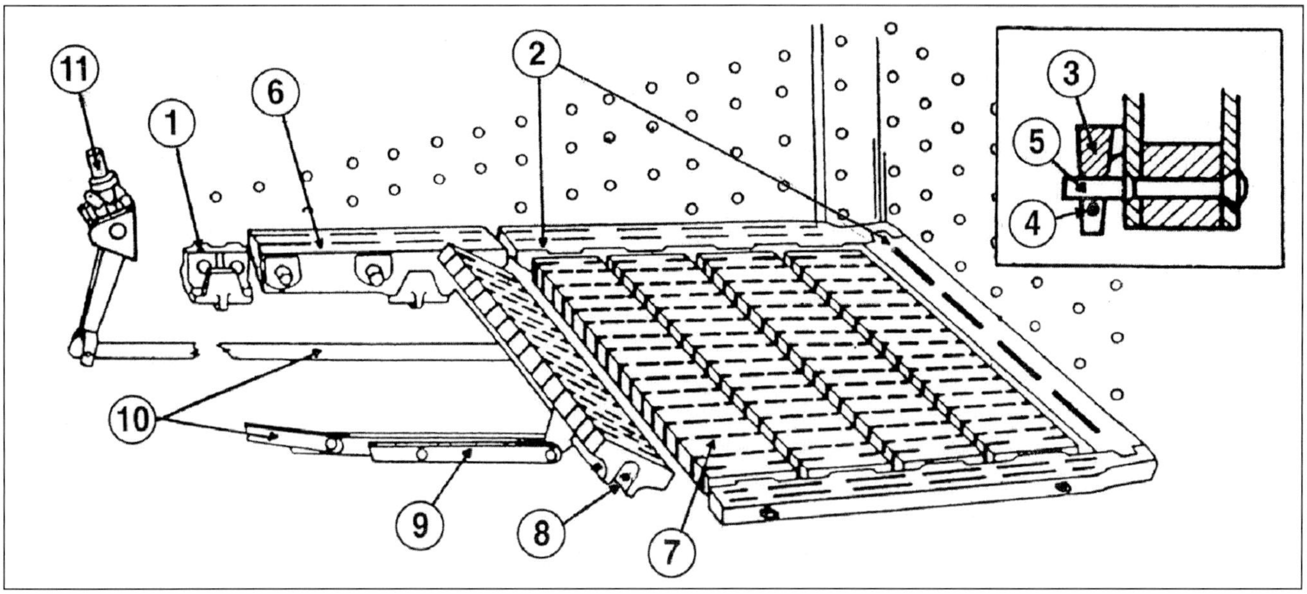

A cut-away view of the rocking grate fitted to LMS and BR Standard locomotives. The rear section has been removed to show the lever mechanism that tilts the firebars. The inset shows the cotter beneath the end firebar: 1 fixing to firebox side; 2 fixed front side and end sections; 3 rear section of firebars; 4 cotter; 5 pin; 6 rear fixed side bars; 7 front section; 8 rear firebar (tilted); 9 firebox linkage to rear section; 10 linkage to front section and operating levers; 11 operating lever spigot

Left: Rocking grate operating lever spigots on the footplate (some utilise a slot). *Brian Dobbs*

Below left: The rocking grate catch lever showing the three lock positions: fully locked; with the outer catch lifted for rocking the fire; and both catches lifted for dropping the fire: 1 extension bar connection; 2 outer catch; 3 inner catch square; 4 guide

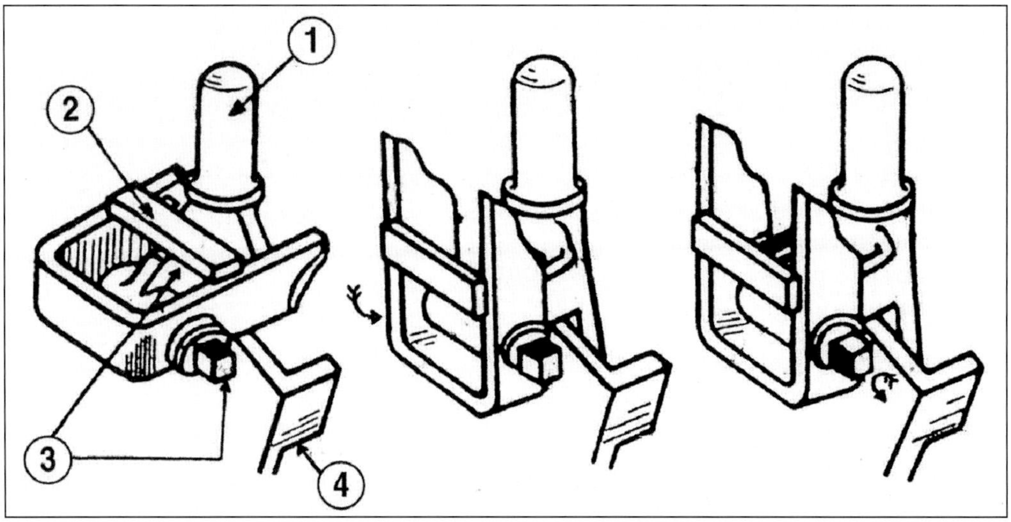

Below: The hopper ashpan, showing the two drop sections and the rods that operate the flaps. The tilting locking clamp can also be seen, with the extension bar attached: 1 damper door; 2 hopper doors;
3 bridge; 4 arm; 5 frame of ashpan; 6 spigot; 7 catch;
8 extension bar

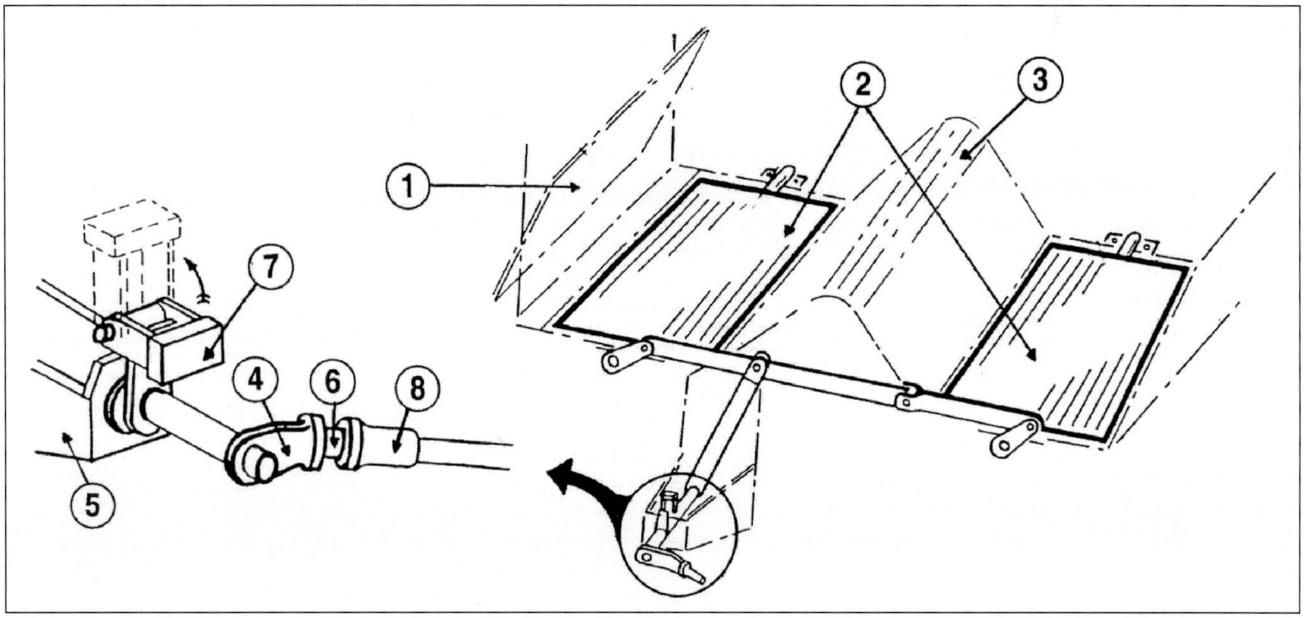

ash and clinker or damage will be caused to the firebars by burning and the fire will be starved of primary air, thus causing bad steaming of the boiler. Before any ash or clinker is raked out of the ashpan it must be soaked with water to prevent it from blowing into the motion parts of the locomotive, which would cause rapid wear of the bearings.

The hopper ashpan

It is a simple matter to empty the hopper ashpan. As with the standard ashpan it is necessary to soak the contents before emptying. The hopper doors are kept closed and locked by a clamp on the left-hand side of the locomotive. This must be lifted and a special bar fitted into the slot to tilt the doors. Some doors are secured by a locking pin, which has to be removed and a handle placed on the square end of the shaft to tilt the drop doors so that the contents can be deposited into the disposal pit. A rapid shaking of the doors will make sure that the hopper is completely empty.

Locomotives fitted with wide fireboxes pitched over the main frames are fitted with flap doors in order to clean the space below the grate. Most of the residue from the fire will fall into the ashpan due to vibration, but the remaining sediment will have to be removed when the fire is cleaned. The best way to clear this space is with a hose pipe, swilling the ash into the main ashpan. Hopper doors and flaps must be closed and secured with the clamp or pin when cleaning has been completed.

Damper doors

Dampers are used to control the heat given from the fire by controlling the primary air to the grate, and they are fitted with doors operated by levers on the footplate. Depending on the size of the ashpan some locomotives have only one damper, usually at the front, but there can be as many as three, as in the case of the Great Western 'Castle' class locomotives.

There are no set rules for the use of dampers, although it is believed that to close the doors when the locomotive is pulling hard will cause the fire to clinker up due to lack of primary air. Only experience will tell. It is recommended that the trailing damper be kept closed when running tender or bunker first as a blow-back could occur otherwise.

CHECKING THE SMOKEBOX

The condition of the smokebox must now be examined, like the firebox, before the fire is lit.

Smokebox doors are hinged and secured by one of two methods. Some former Southern Railway and early LMS locomotives, such as the 'Jinty' 0-6-0Ts, 4F 0-6-0s and 'Crab' 2-6-0s, have their doors secured by a series of bridge clamps with single nut fixings. Most other locomotives have a single centre bolt fixing called a dart, because of its shape. The dart is fitted with two handles, or one handle and a wheel. The inner handle in both examples controls the rotation and location of the dart. The handle should be in line with the lug at the end

The smokebox door of an Ivatt Class 2 locomotive, showing the two handles. The inner handle locates the dart, while the outer handle tightens the door. *Brian Dobbs*

Left: The horizontal slotted bar for locating the lug at the end of the dart. The pin that secures the bar can be seen on the left; this can be removed to allow the bar to swing out to facilitate the cleaning out of ash. Note the jet ring on top of the blastpipe. The locomotive is 'Black Five' No 5407. *D. Dyson*

Below left: No 71000 Duke of Gloucester has a vertical centre bar in the smokebox, although the slot is still horizontal. The adoption of a vertical bar was a development on BR Standard locomotives that made the emptying of the smokebox an easier task as it eliminated the need to remove the bar in order to gain access. *D. Dyson*

Below: The single nut and bridge clamp smokebox door fitting. As can be seen, one side of the clamp fits over the edge of the door while the other leg is firmly bolted on the edge of the smokebox. *Author*

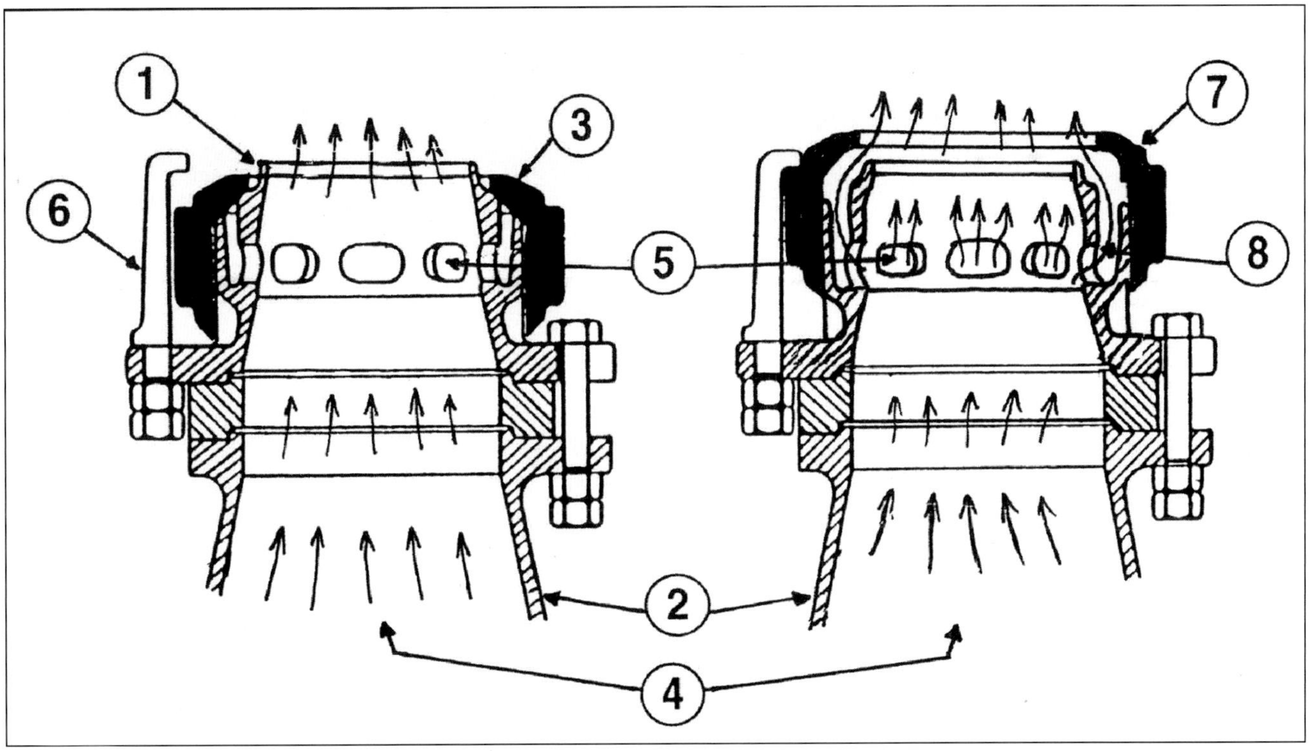

The Great Western 'jumper ring' fitted to the blastpipe. That on the right is open, as when the locomotive is working hard, while the one on the left is closed: 1 blastpipe cap; 2 blastpipe; 3 jumper ring in closed position; 4 direction of steam from cylinders; 5 ports to blastpipe; 6 stop; 7 jumper ring in open position; 8 additional outlet for exhaust

of the dart, which engages in the central slot of a removable bar extending across the inside of the smokebox. When the handle is turned through 90 degrees the lug will be held behind the slot in the bar. The outer handle, or wheel, has a threaded centre which is screwed on to the dart to secure the door in an airtight condition.

To open the door turn the outer handle anti-clockwise until the inner handle can be turned through 90 degrees to the horizontal position. This releases the lug of the dart in the slotted bar and the door can then be swung open.

The early Midland type of smokebox door is fastened with single-nut bridge clamps. Simply slacken each nut with the correct size of spanner until the bridge clamp can be turned to clear the rim of the smoke box door.

With the door open, a detailed examination of the smokebox can be carried out. Just as with the firebox, the tubes and washout plugs must be examined for evidence of water leaks. Any ash left over from the last time the locomotive was in steam must be removed so as to give a clear passage of air through the tubes.

Some Great Western engines have a 'jumper ring' fitted to the top of the blastpipe, which is lifted by the blast of the escaping steam up the blastpipe. Its purpose is to alter the choke on the blastpipe and lessen the blast effect of the escaping steam on the fire when the locomotive is working hard. This must be checked for free vertical movement, as carbon and ash may have caused it to become seized.

The smokebox door must be airtight when closed, otherwise ash that accumulates in the smokebox may be ignited by the exhaust gases and burn, causing damage to the smokebox door and tubeplate. Before closing the door any ash must be removed from around the rim as this could possibly effect the airtight seal.

If air is being drawn into the smokebox from a defective door seal, this can also cause poor combustion of the fire and seriously affect the steaming quality of the locomotive as it will reduce the effect of the blast and thus the amount of primary air being drawn through the firegrate and along the tubes.

Before the Midland-type door can be closed the clamps must be clear of the rim. The door is then pushed closed and the clamps are turned so that the long leg is against the rim of the smokebox and the short end is over the rim of the door. When the nuts are tightened the clamps will secure the door and ensure that the smokebox is airtight. A check can be made when the locomotive is in steam and the door is secured by opening the blower valve and listening for air leaks at the rim of the door.

LNER streamlined A4 class locomotives have a steel visor fitted over the smokebox door that must be opened to gain access to the smokebox. This is done by means of a screw mechanism on the left-hand side of the locomotive, just in front of the cylinders.

LIGHTING THE FIRE

When all the above checks have been completed and everything is confirmed as being in a satisfactory condition, a start can be made on lighting the fire.

Fire-lighting methods

There are three methods or places where a fire can be lit in the firebox, and each has its own merits. It is important that steam is not raised too quickly because the different expansion rates of the metals in the firebox could cause the tube ends to leak or the stays to break.

A large locomotive can take up to eight hours to raise steam, so it is advisable to light a small fire in the grate the day before in order to warm the water around the firebox and in the boiler. The fire can then be left to burn out. There must of course be sufficient water in the boiler.

One way of lighting the fire is to ignite it directly under the firebox door, as was the practice in the British Railways era. However, this method, although simple and easy, is not very satisfactory because the smoke given off by the fire will drift on to the footplate when the door is opened, there being no steam in the boiler to operate the blower valve. However, this problem can be partly eliminated by removing the baffle plate.

The second and most efficient way to light the fire is under the brick arch. By this means most of the smoke and heat will be retained inside the firebox. Start by placing a few shovels-full of small coal, about the size of a man's fist, in the square under the brick arch to a depth of about 4 inches but not in contact with the sides of the firebox (to prevent localised heat should the fire start to burn in one area, which would affect the plates before warming the firebox, thereby causing damage to the stays).

To light the fire place two or three domestic firelighters or some rag soaked with a mix of oil and paraffin (such as that used for cleaning) together with some pieces of dry wood on to the coal under the brick arch. Next place an amount of rag soaked with paraffin on the paddle, ignite it just inside the firebox door and lower it on to the firelighters or the pile of rags and wood under the brick arch. It will be necessary to leave the damper door closed until the fire is burning well in order to keep the smoke away from the footplate.

When the wood and coal is well alight, more small coals can be placed on to the burning fire. The damper can now be opened and the fire can be left to burn through so that heat is built up gradually in the firebox. As time goes by the fire can be built up sufficiently to spread all over the grate in order to raise steam once the firebox has been warmed through.

These first two methods are mainly used when the locomotive is being lit up for the following day. The third method is to coal the entire firebox all over the grate to a depth of about 4 inches, then, by the use of fire-lighters or rag and wood, to light a fire in two or more areas, depending on the size of the grate.

Before the fire is lit, any ash or clinker must be removed from the grate either by lifting out the firebars or digging it out through the firebox door with a steel shovel called a paddle. *All photographs by the author*

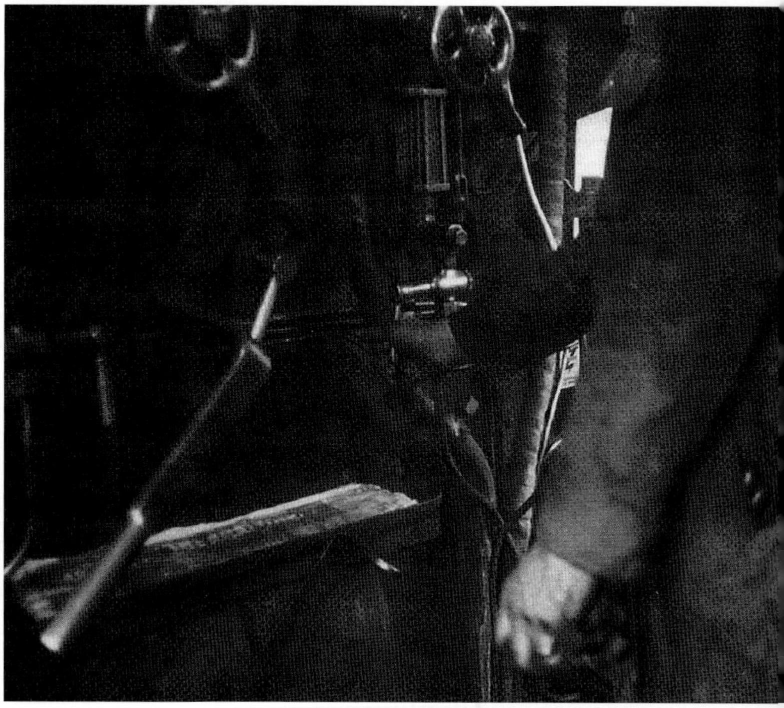

Check that there is water in the boiler by operating the gauge frame.

CHECKING THE LUBRICATORS

If the locomotive is fitted with a hydrostatic lubricator it is generally assumed that it is the fireman's duty to replenish it with oil. To do this the water must first be drained from the lubricator body. If the locomotive is in steam it needs to be ascertained that the steam supply from the boiler to the lubricator has been shut off.

With former Great Western locomotives the steam shut-off cock has a pointer and is fitted between the steam manifold and the condensing coil in the cab roof. This cock must be in the closed position. On top of the lubricator body (see also Chapter 1) is a two-way valve that must also be in the closed position. On the bottom of the body is the drain tap. When this is opened water will be seen running away down the drain. It is always important to open this drain cock before the hexagon

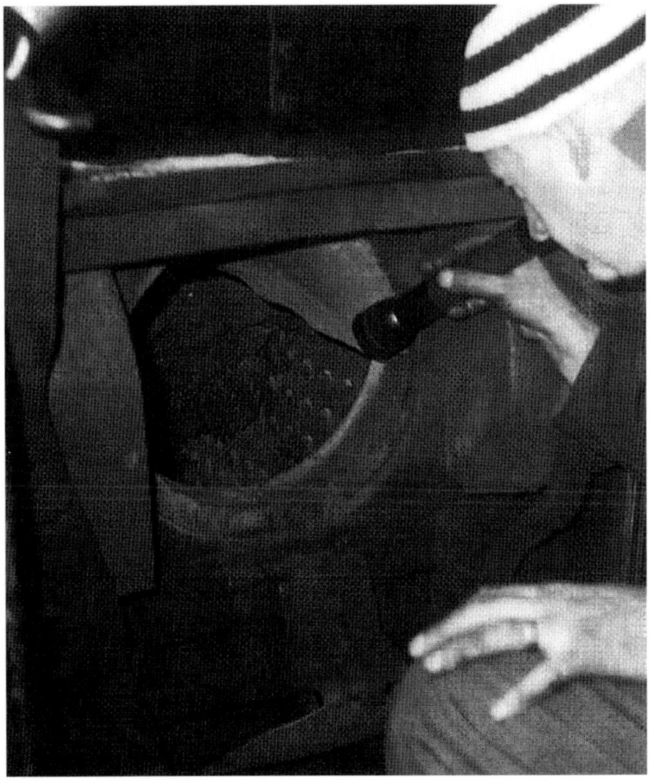

Check the inside of the firebox for cleanliness and water leaks, and the condition of the firebars, brick arch and fuseable plugs.

Coal part of the firegrate to a thickness of about 4 inches; do not place coals near the firebox walls. Next place dry wood on the coal followed by oily rags such as those that have been used for cleaning.

Check the smokebox. First slacken the outer handle of the smokebox door, turning it anti-clockwise.

Once the tension is slack enough to do so, turn the inner handle to the horizontal position; this will move the dart inside the smokebox to enable it to pass through the slot in the bar and the door will open.

Pull open the door and check that all the ash has been removed.

If all is clear close the door, making sure that the dart has passed through the slot in the bar, then turn the inner handle through 90 degrees to the down position. Tighten the outer handle, locking the door and keeping it airtight.

Next check the ashpan and empty it if necessary. To empty a hopper ashpan, first soak the contents to save the dust from blowing all over the locomotive. Next remove the catch (as described earlier in this chapter) and use the extension bar to open the hopper doors. Make sure all the contents are removed by repeatedly opening and closing the hopper doors. Early types of ashpan have to be checked by opening the damper and going under the locomotive to inspect the pan.

Before lighting the fire make sure that everything is closed, such as the regulator, blower valve, ejectors, etc., but leave the damper down. Place some domestic firelighters on the wood and coal and ignite them using a piece of rag soaked in paraffin on a shovel or paddle.

filling nut on top of the lubricator body is removed as this action will release any pressure that may be left in the system.

Using a special Great Western spanner, slowly remove the hexagon filler nut. Take care to stand back and do not lean directly over the lubricator if the locomotive is in steam, just in case there is any pressure remaining. Leave the water to drain away until the oil can be seen flowing out at the bottom.

When the water in the lubricator has run out of the body the bottom cock can be closed and the lubricator refilled with thick cylinder oil up to the base of the thread in the filler orifice. The nut is replaced and tightened with the spanner.

The hydrostatic lubricator fitted to GWR locomotives can have three, four or five sight glasses through which the oil can be seen passing, so that the flow can be adjusted by the driver. Sometimes the inside of these glass tubes needs to be cleaned. To do this, first remove the top nut directly over the glass tube and, with a piece of clean rag on the end of a suitable screwdriver, clean out the tube.

After cleaning, a very small drop of washing-up liquid or ordinary antifreeze, about the size of a pin head, is placed in every tube; this will keep the glass clear for some time. Refill the tubes with clean water, replace the top nuts - not forgetting the copper sealing washers - and tighten the nuts. If the locomotive is in steam when the lubricator has been filled, the warming cock should be opened to warm the oil in the body.

Some industrial locomotives have a smaller but similar type of hydrostatic lubricator and these are drained and re-filled in a similar way to the GWR pattern, although they may not be equipped with a warming device.

FILLING THE TANKS OR TENDER

Before leaving the shed the water level in the tanks or tender must be checked and replenished if necessary. It is important when filling that the water is not allowed to overflow from the tank, especially with a tank engine, because it can find its way into the axleboxes. While the tank is filling with water it provides a good opportunity to make sure that the coal and fire irons are stacked safely on the tender or bunker so that they will not fall off on to anyone below or become lost along the trackside.

CHECKING HEADLAMPS

While steam is being raised the engine lamps must be attended to. There are three designs of headlamp in use today. The first is the round type with a hinged lid retained by a clip that forms the ventilator. The second is the square type with the door at the side to g ain access to the cistern and a ventilator on top of the body. Both these lamps have a mounting at the rear to accept the locomotive lamp bracket and are interchangeable. The third type is the Great Western lamp, which has a side-mounted fitting and therefore can only be attached to GWR locomotives.

The lamps should be cleaned and fitted with one red shade per lamp. The cistern needs to be cleaned and the wicks checked for condition. The wicks must not be hard to the touch or they will not syphon the paraffin; they must also be of sufficient length to reach the oil in the cistern. Turn the adjusting wheel on the side of the burner to bring the wick out

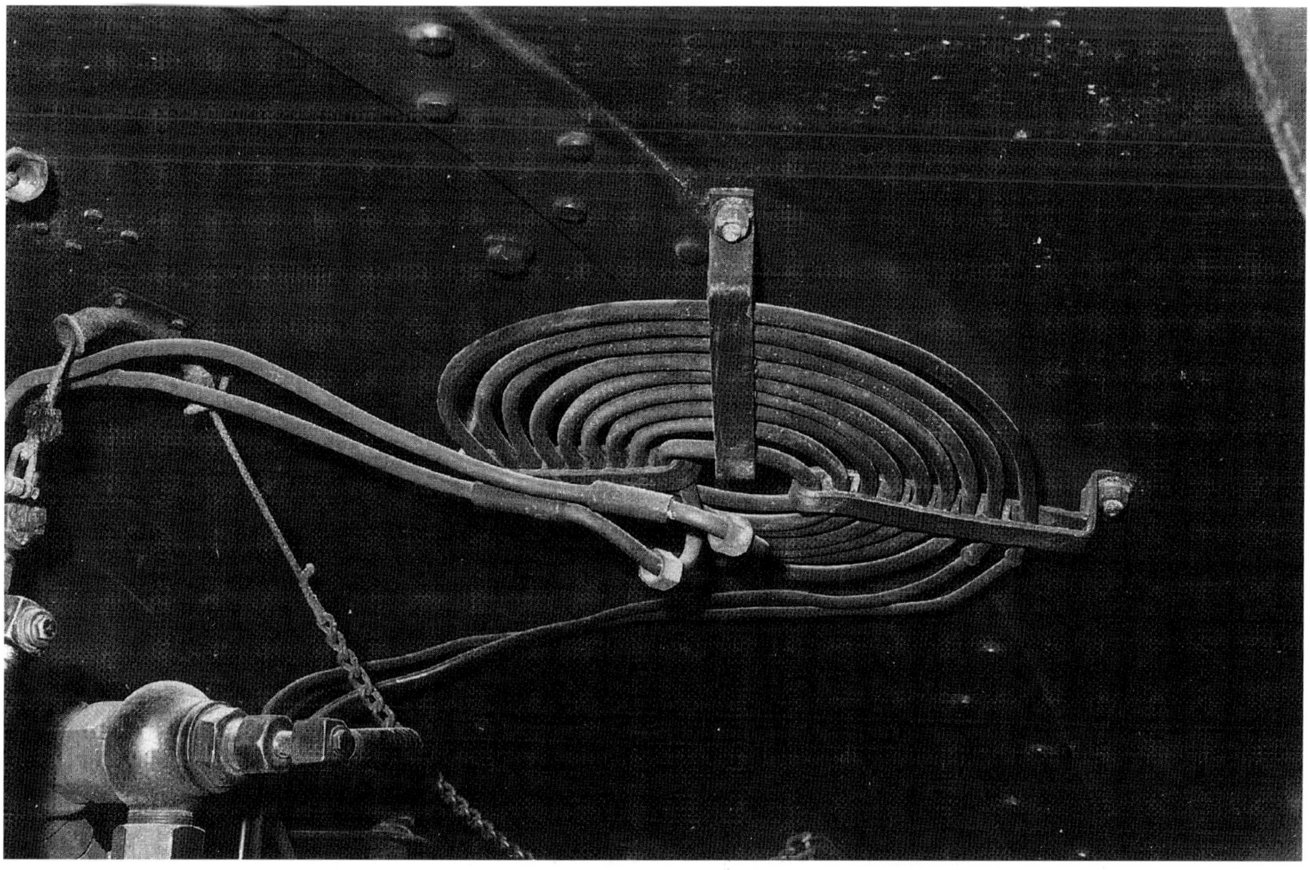

The condensing coil fitted to the cab roof of a former Great Western 'Manor' class locomotive. *D. Dyson*

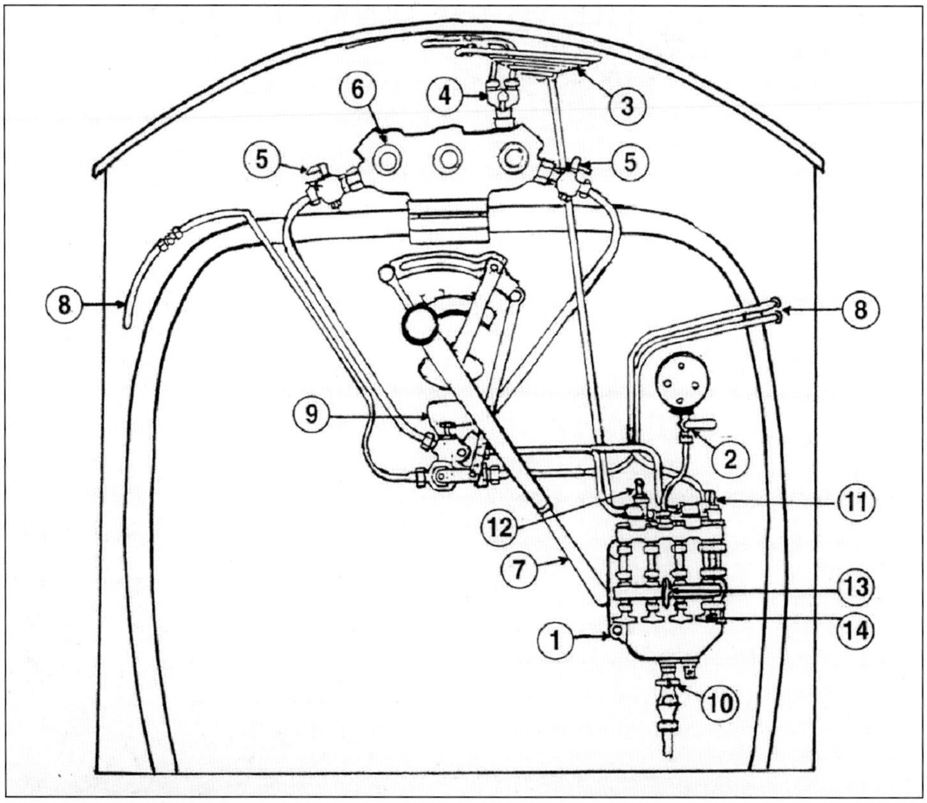

A sketch of the lubrication system inside the cab of a GWR locomotive: 1 lubricator body; 2 warming cock; 3 condensing coil; 4 steam cock (up for closed, left for left coil, right for right coil, and down for both); 5 oil and steam isolating cocks; 6 steam manifold; 7 regulator handle; 8 steam and oil pipes; 9 jigger oil supply valve; 10 drain; 11 filler nut; 12 isolating cock; 13 main gallery shut-off; 14 feed regulators

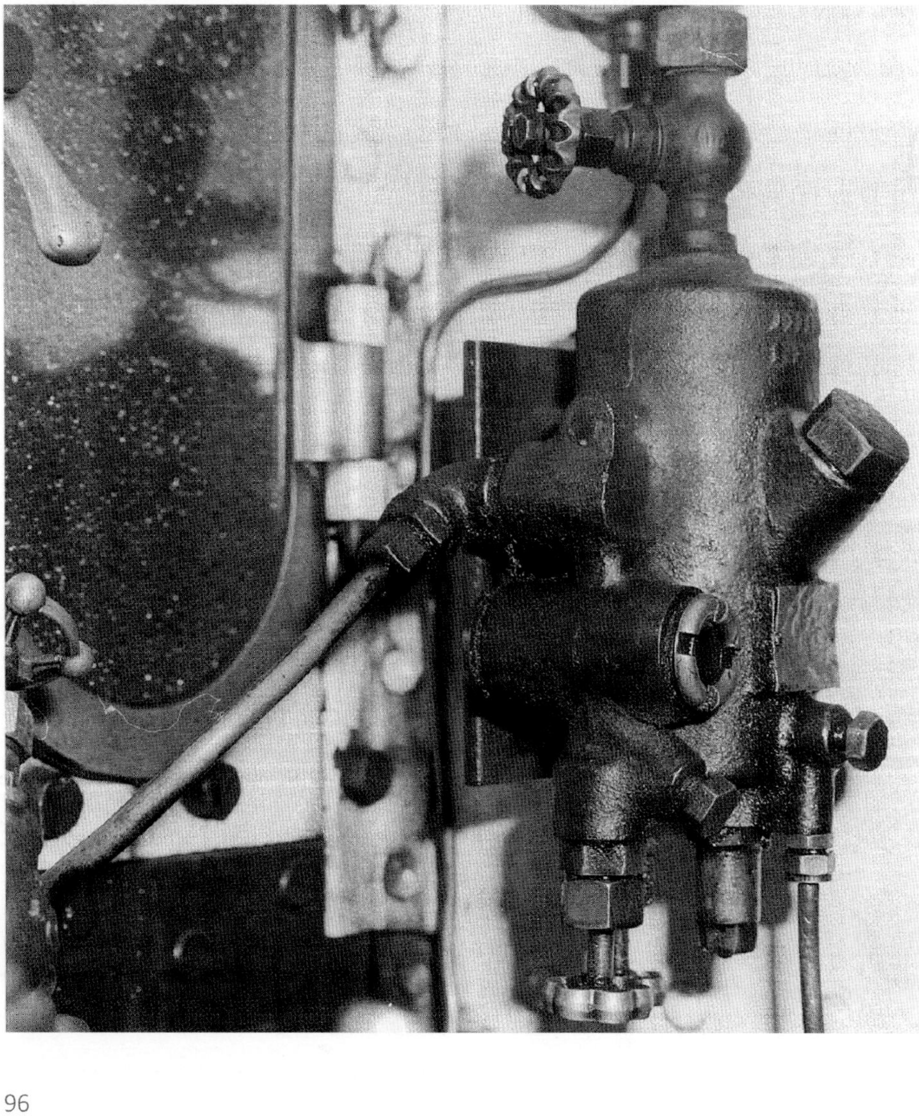

A typical lubricator fitted to an industrial locomotive. At the top is the steam shut-off wheel, while below it the large nut pointing upwards at an angle is the filler. The smaller nut is to drain the water from the main body. On the bottom is the wheel that adjusts the flow of oil. *Brian Dobbs*

Filling the Great Western lubricator. First close the main steam supply to the lubricator at the manifold, then open the drain nut at the base of the lubricator body and slacken and remove the filler nut at the top. When all the water has drained from the lubricator, the drain can be closed and tightened. The lubricator can then be refilled with steam oil to the base of the thread in the filler orifice. The job is completed by replacing the filler nut and turning on the steam at the manifold. The oil will have to be warmed before use, and this is done by opening the lubricator warming cock, allowing steam through the lubricator body. *R. G. Fox*

Filling the tender tank from a platform water column. While on the tender check that the coal and fire irons are stacked safely. *Brian Dobbs*

Both types of square headlamps, the side fitting type of the Great Western and the rear mounting type.

To clean, fill and trim the lamp, first open the side door and remove the cistern assembly from the lamp body.

Tilt back the glass-fronted draught protector and remove the burner from the cistern. Clean the lamp body, not forgetting the bull's-eye front lens, and the cistern, then refill with paraffin to the correct level; do not overfill as the heat can cause the fluid to expand and spill, setting fire to the lamp. A good guide is when the fluid is level with the two holes in the recess for the burner. Also examine the wick. The fabric should be long enough to reach the fluid and not be hard to the touch. At the top of the wick will be found a hard piece of carbon formed during previous use. This must be removed before lighting.

Above: Light and adjust the burner to give the optimum flame. The flame should not be so high as to smoke and deposit carbon on the ventilator on top of the lamp body. Neither should it be so low that the lamp will blow out.

Left: Replace the cistern into the slides on the base of the lamp body. It may be necessary to leave the flame high until it is surrounded by the wind protector, to be adjusted when the assembly is in the lamp body. Here the front door is open showing the correct flame length. The red aspect is seen resting on top of the slide; when this is in position between the burner and the lens the lamp aspect will be showing red for use as a tail lamp.

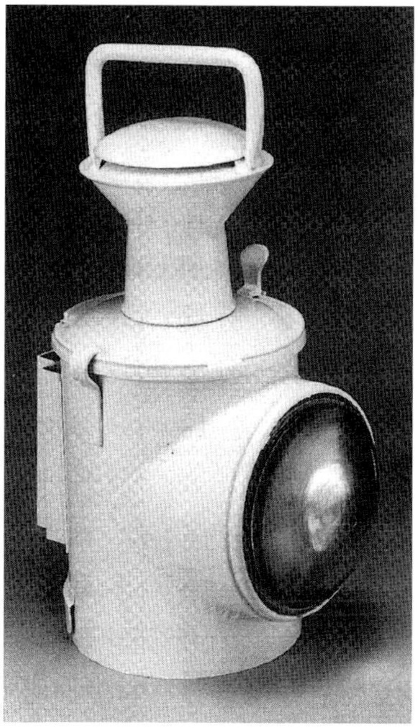

Access to the burner of the round-pattern lamp is through the removeable centre section of the lamp, which contains the red and clear shades. The lamp should be cleaned and refilled in the same manner as with the square pattern. The cistern is retained by a clip fitted round a bracket at the rear of the lamp body. The colour aspects are changed by rotating the centre section, each position being secured by a spring-loaded lug. Another type of round lamp has a hinged lid. *Beesley Products*

sufficiently from the top of the body. At the top will be found a burnt piece of carbon, which must be removed by pinching between the finger and thumb before the lamp is lit. After filling the cistern, replace the burner and light the lamp, then adjust the flame until there is no trace of smoke, otherwise the lamp will carbon up. Neither must the flame be so low as to be blown out by the motion of the locomotive.

STEAM-RAISING

Constituents of coal

Coal consists of six basic elements that can be remembered easily by means of the acronym 'NO CASH', which stands for Nitrogen, Oxygen, Carbon, Ash, Sulphur and Hydrogen (see the accompanying graph).

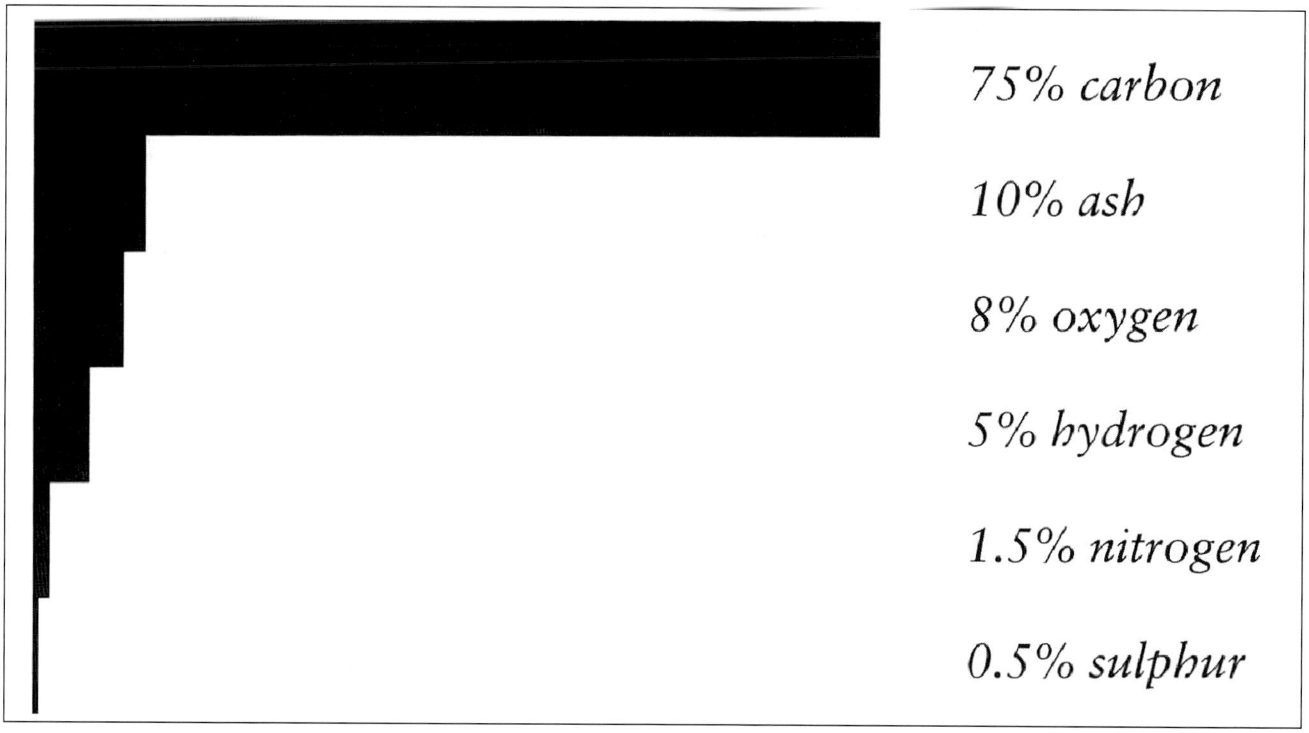

A graph showing the constituents of typical steam coal, carbon being the main one.

When coal is burned in the firebox of a steam locomotive it gives off several gases that will turn into heat. Complete combustion will not take place until a temperature of about 2,500°F has been reached. Coal will burn at a lower temperature of about 800°F, but will give off carbon monoxide gas, which is poisonous and will be wasted, not being ignited until the fire has produced a temperature in excess of 1,200°F. With the complete combustion of coal in the firebox the carbon monoxide gas will turn into carbon dioxide, and will then burn completely in the combustion chamber.

A common cause of low firebox temperature is excessive coal, which prevents sufficient air from reaching the firebed. As explained earlier, in order to produce heat and make steam two types of air are admitted to the fire. Primary air is admitted to the grate through the damper door of the ashpan, and causes the greater combustion of the coal and increased heat in the firebox. Secondary air is admitted by opening the firebox doors; this passes directly over the fire and does not create a great amount of heat, but assists the combustion of the carbon gases given off by the burning coal.

While building up the fire, attention must be given to the smoke being emitted. Black smoke can and must be kept to a minimum by adjusting the firebox doors and making use of the blower valve if the locomotive is in steam. If there is green sulphurous smoke at the chimney top, this indicates that there is so much coal on the fire that insufficient oxygen is getting to the grate, resulting in incomplete combustion of the fuel.

Again, this problem can be avoided by opening the firebox doors and by the use of the blower valve.

Blower valve

The blower valve, sometimes called the 'jet', is used to create a partial vacuum in the smokebox. It is operated by steam and controlled from a valve on the footplate, and admits steam at boiler pressure through a series of holes in a casting that sometimes forms part of the blastpipe and is directly in line with the petticoat pipe that forms the base of the chimney. The jet ring must not be confused with the exhaust pipe from the vacuum ejector, which also feeds into the chimney, usually at the base of the petticoat pipe.

As the steam is admitted to the casting it causes air to be expelled from the smokebox and up through the chimney, thus creating a partial vacuum in the smokebox. As the smokebox is airtight this causes air to be drawn from the firebox along the tubes and out to the atmosphere via the chimney. The purpose of the jet is to prevent the risk of a blow-back of flame through the fire doors, and to control the smoke given off by the fire. It can also be used to raise steam quickly by increasing the flow of air through or over the firebed.

When the locomotive is in steam the blower valve must always be open to prevent smoke from the fire covering the faceplate. It should be opened before the regulator is closed to prevent incomplete combustion of the coal in the firebox and to prevent the gases from escaping through the firebox doors engulfing the footplate in flame. It is essential that the blower is increased before passing through tunnels, under bridges and through deep cuttings.

Raising the steam pressure

When it is considered that the firebox is warm enough, the fire can be spread all over the grate in order to raise the steam pressure. Certain types of coal will make clinker quickly, and this will clog the firebars, causing the fire to be starved of primary air and making the locomotive hard to steam. This problem can be alleviated by spreading limestone or pieces of broken fire-brick on the grate before the fire is spread so that the clinker cannot form a complete sheet over the grate area.

Other types of coal will make large quantities of ash, which will also clog the grate. This problem can be eliminated by removing one or two firebars from each row before the fire is spread. This will create a greater gap between the bars, allowing the ash to fall through into the ashpan.

Only the recent experience of others will be of assistance in helping with what action to take, if any, as they will know the quality of the coal that is being supplied for use. In the heyday of steam railways coal was supplied regularly from particular collieries, such as soft Welsh or hard Yorkshire coal, and the action required by the firemen was known and the fire bars were designed accordingly.

OPERATING THE INJECTORS

When the fire has been spread and sufficient steam has been raised it is time to test the two injectors, whose purpose is to feed water into the boiler. Where both injectors are live steam they can be tested by the fireman, but the driver's assistance will be required when testing an exhaust steam injector.

A description of the design and types of injector has already been given in Chapter 1. Usually there are two injectors fitted to a locomotive, and it is advised that they are operated alternately; indeed, it is now a legal requirement to have two independent means of feeding water to the boiler. In the past some locomotives had water pumps driven from the motion, but these are not used today.

Live steam injector

First check that no one is standing near the water overflow, then open the water handle until water can be seen at the overflow. Next open the steam valve, or spindle, about half a turn and pause for 5 seconds. Steam will then be seen emitting from the overflow mixed with water. The steam spindle can then be fully opened to allow steam to enter the injector. Next the water flow must be adjusted or trimmed by slowly closing the water handle to allow condensation of the steam to take place in the combining cone and to overcome the boiler pressure. If the water handle is closed too much the injector will blow back, and steam will be seen blowing violently from the overflow. If this happens, quickly open the water handle and try again to trim the injector by closing the water handle more slowly. It may be necessary to close the steam spindle and start again. The injector can be heard and seen to be working efficiently when no water is running out of the overflow.

When the required boiler water level has been achieved the injector can be closed down by the following method. First close the steam spindle fully, then close the water

A standard faceplate-mounted or lifting injector fitted to a tank locomotive. The water handle is in the horizontal, closed position and the steam spindle is controlled by the wheel. The square nut to close off the top clack can be seen just behind the steam spindle. *Author*

handle, making sure that no water or steam can be seen at the overflow. If steam is allowed to drift through the injector body it will warm the condensing cone and make the injector difficult to re-start. If the steam spindle is closed tightly and steam continues to emit from the overflow it must be coming from the top clack, indicating that it has not seated properly or is partially stuck. The injector must then be re started again for a few seconds then the steam spindle closed quickly in an attempt to cause the top clack to reseat itself.

Faceplate-mounted (or lifting) injector

This type of injector is found on older types of locomotive and is sometimes called the lifting injector. Where the injector is placed under the footplate steps gravity will fill it with water when the water valve is opened, but the faceplate-mounted injector is usually higher than the water supply.

The method of operation is as in any other live steam injector: open the water valve to allow the water to flow, then open the steam spindle slightly in order to lift the water to the body of the injector. After 5 seconds fully open the steam spindle and adjust or trim the water handle to start the injector fully. The injector can be closed off in the normal way by first closing the steam spindle, then the water handle. An inherent fault with the faceplate-mounted injector is that it gets too warm, which stops condensation taking place in the combining cone, causing difficulty in starting. This problem can be overcome, however, by pouring cold water over the body of the injector to cool down the combining cone.

Exhaust steam injectors

There are many types of exhaust steam injector fitted to steam locomotives in use today, although they all work on the same basic principle, utilising exhaust steam that would otherwise be wasted up the chimney. Exhaust steam injectors are capable of working on both live and exhaust steam, and also provide a supply of hot feed water to the boiler, thereby providing a saving in the overall coal consumption of the locomotive.

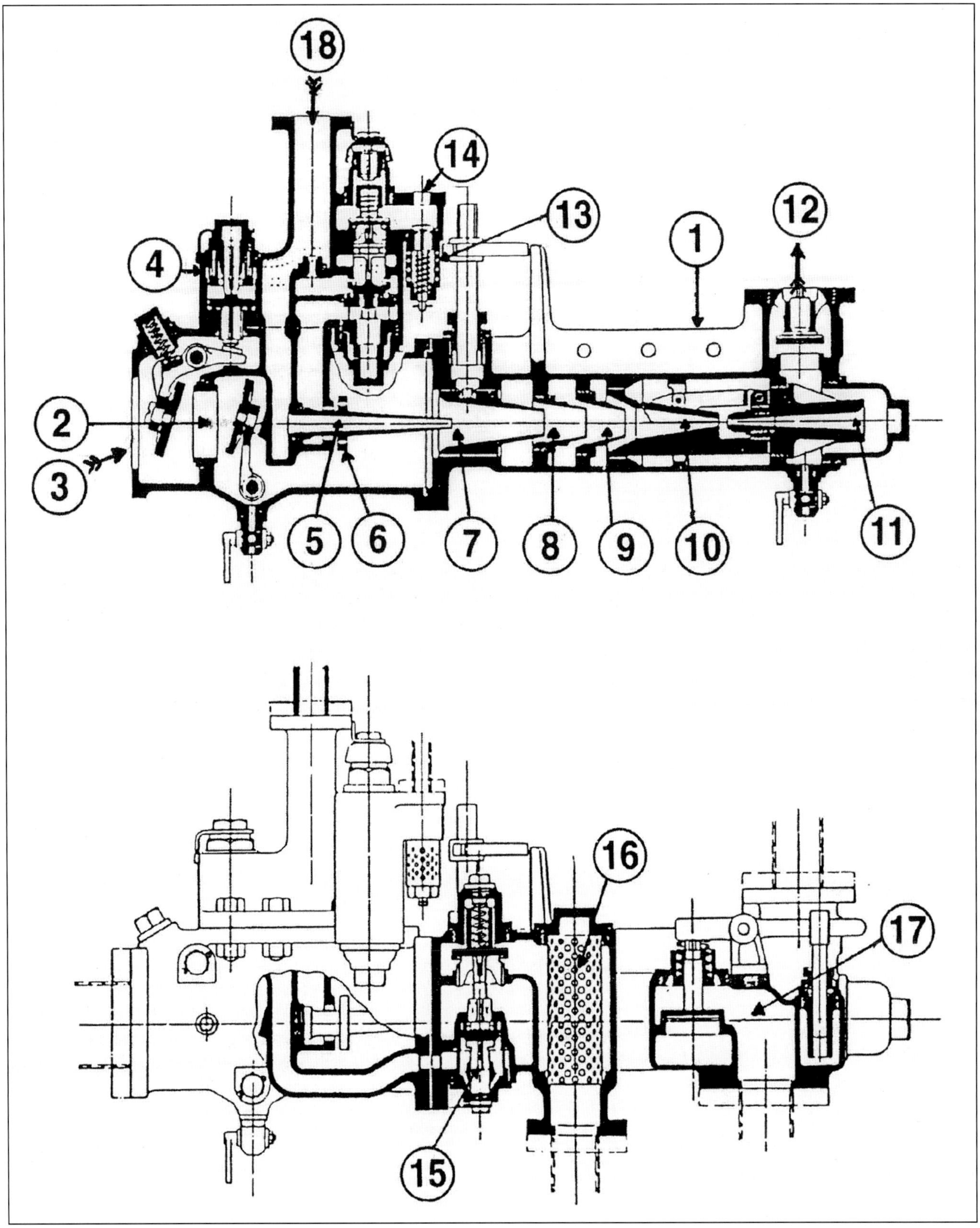

The Class H exhaust steam injector. The upper view shows the arrangement of the cones. Note the separate live and exhaust steam valves. The latter (2) has a spring-controlled outer valve, while the inner valve is live-steam-controlled. The lower view shows the water control valve and the automatic water overflow valve on the far right, which is kept closed by water pressure: 1 injector body; 2 exhaust steam valve; 3 exhaust steam supply; 4 exhaust steam control piston; 5 live steam cone; 6 auxiliary live steam nozzle; 7 exhaust steam cone; 8 draught tube; 9 vacuum tube; 10 combining cone with hinged flap; 11 delivery cone; 12 delivery; 13 drip valve; 14 control pipe from engine steamchest; 15 automatic water control valve; 16 strainer; 17 automatic overflow valve; 18 live steam supply

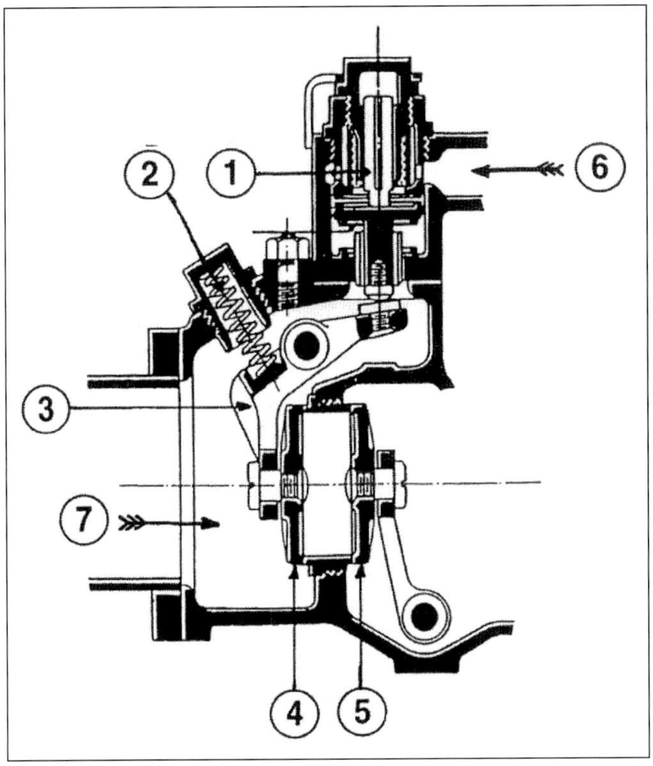

Detailed diagram of the automatic exhaust steam valve: 1 exhaust steam control piston; 2 spring; 3 crank lever; 4 outer exhaust steam valve; 5 inner exhaust steam valve; 6 live steam supply; 7 exhaust steam supply

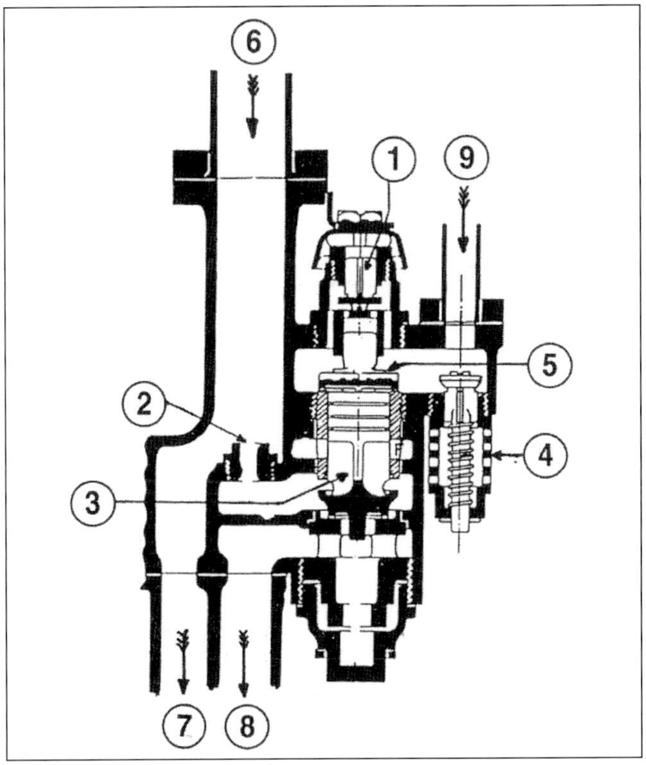

Detailed diagram of the automatic shuttle valve and change-over mechanism: 1 anti-vacuum valve; 2 choke; 3 automatic shuttle valve; 4 drip valve; 5 auxiliary check valve; 6 live steam supply; 7 to live steam cone; 8 to auxiliary live steam nozzle

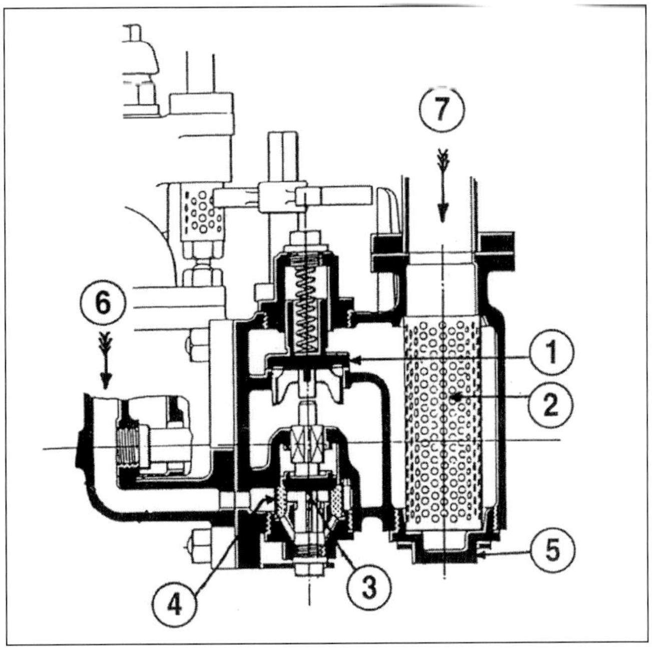

Detailed diagram of the automatic water control valve. Note the live steam supply from the boiler that lifts the control piston of the water valve, allowing water to enter the injector: 1 water valve; 2 strainer; 3 control piston; 4 live steam strainer; 5 strainer cap nut; 6 live steam supply; 7 water supply

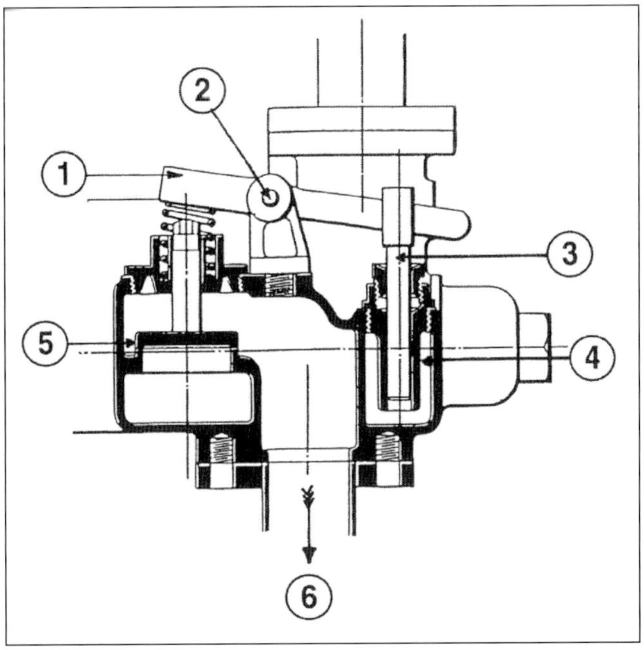

Detailed diagram of the automatic overflow valve: 1 lever; 2 fulcrum; 3 water-controlled piston; 4 cylinder; 5 overflow valve; 6 overflow

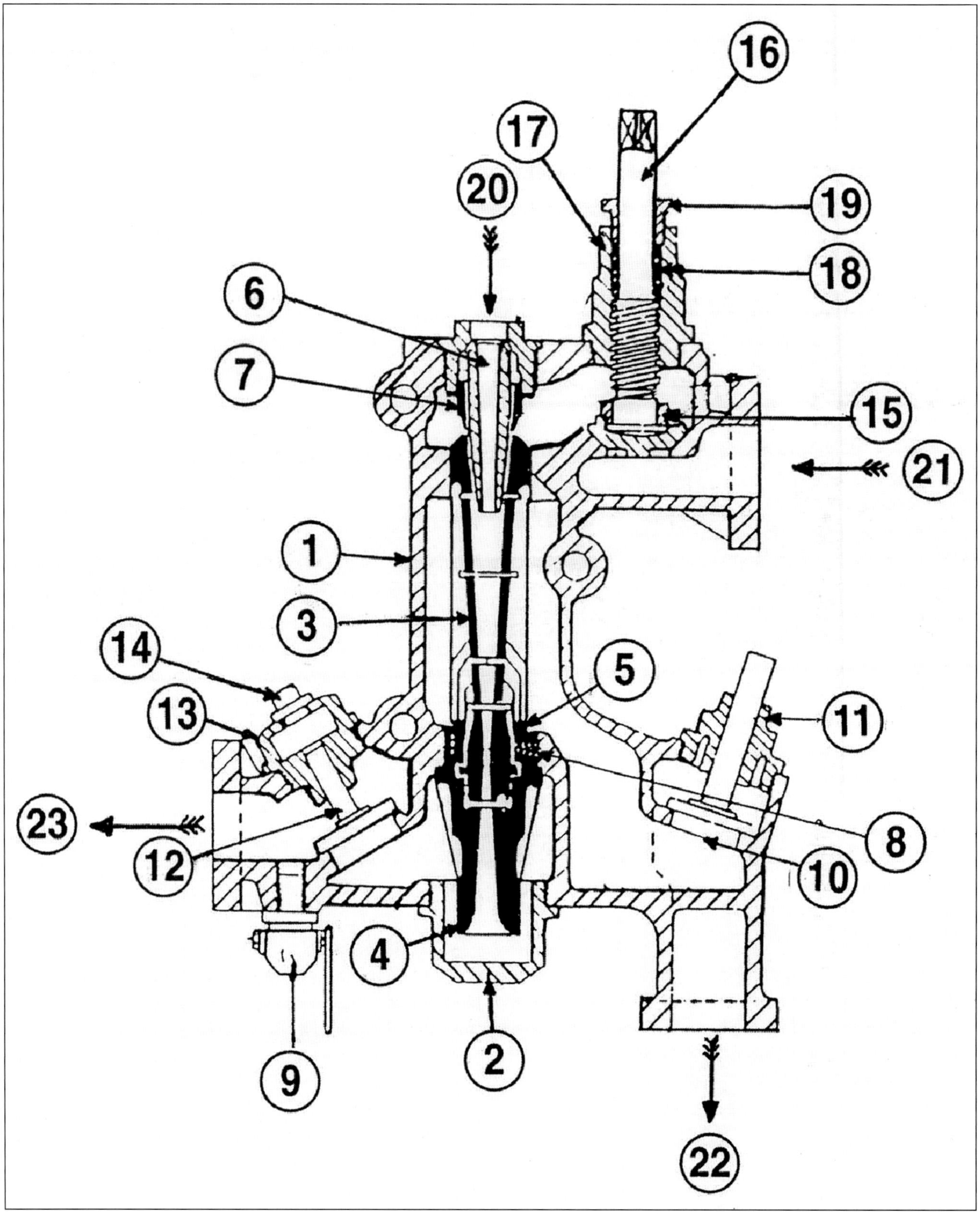

The monitor injector: 1 injector body; 2 delivery cap nut; 3 combining cone; 4 delivery cone; 5 renewable end; 6 inner steam cone; 7 outer steam cone; 8 screw; 9 drain cock; 10 overflow; 11 overflow cap nut; 12 back pressure valve; 13 back pressure valve cap nut; 14 grinding nut; 15 water valve and nut; 16 water valve spindle; 17 box nut; 18 neck ring; 19 packing gland; 20 steam inlet; 21 water inlet; 22 overflow; 23 delivery

A monitor injector fitted under the footplate of a locomotive. Refer to the accompanying diagram to identify the various parts. *D. Dyson*

Exhaust steam injector (Class H): The water valve is controlled by the admission of steam to the injector, and the method of operation is simple (there is a separate water control handle that closes off the water supply to the injector, but this can be left open). Open the steam spindle slightly, about 90 degrees, and wait for about 5 seconds while live steam is fed from the boiler to operate the steam-controlled water valve in the injector body. This will allow water to enter the injector through a strainer, and water will be seen running from the overflow. The steam spindle can then be fully opened and the injector started. It may still be necessary to trim the injector by operating the water regulating control handle.

It was found that there was a need to fit a safe and positive water overflow valve and this works as follows. A piston fitted in the delivery chamber of the injector is connected to a pivoting lever, the other end of which bears against the upper stem of the overflow valve. When the injector is working the delivery water pressure under the piston forces up the lever, which holds the overflow valve on to its seat. If the injector stops working, the pressure under the piston immediately falls, allowing steam and water to escape through the overflow until the injector re-starts. The pressure under the piston then re closes the valve.

The action of opening the live steam valve on the footplate will operate the control piston of the automatic exhaust steam valve in the injector body. This will apply pressure to an arm controlling the outer exhaust valve, which is retained on its seat by a spring. If the locomotive is working there will be a supply of exhaust steam for the injector taken from the base of the blastpipe in the smokebox. The steam passes to the injector via a grease separator fitted to the smokebox steam supply pipe between the blastpipe and the injector. This works automatically and requires no participation by the locomotive crew. The particles of grease that are found in the exhaust steam originate from the lubricant supplied to the cylinders, and must be separated or they will cause the boiler to become greasy and prime.

If the regulator is closed and no exhaust steam is available, the automatic exhaust steam valve will also close and the injector will return to live steam operation.

When the injector is required to work with live steam only, it operates in the following way. When live steam is admitted to the injector by opening the steam valve on the footplate, this will release steam to the automatic shuttle valve. The inner exhaust valve will be kept closed in the body of the exhaust control valve by the pressure in the valve body, allowing the injector to work on live steam only. When the regulator is closed the pressure on the steam-controlled water valve and the automatic exhaust steam valve will also be closed automatically, as the steam supply has been cut off.

The supply of water to the boiler is controlled by a water regulating handle on the fireman's side of the locomotive. It is advisable not to use the exhaust steam injector while shunting or working where the driver's steam regulator is being opened and closed constantly. This is because the shuttle valve that changes the steam supply from exhaust to live steam will be constantly operating and this may cause the injector to blow back.

Water must never be allowed to waste from the injector overflow as this could soon empty the tank, leaving the locomotive short of water.

Exhaust steam injector (Class J): This works in the same way as the earlier classes of exhaust injectors, the main difference being the exhaust steam supply valve. Instead of having two flap valves to control the steam, the J type has a bobbin valve controlled by live steam from the manifold.

Monitor injector

This type of injector was one of the latest developed for steam locomotives and utilised live steam only. It was fitted with the sliding-type combining cone to relieve pressure should the m1ector blow back. Vertically mounted, it can be found fitted under the steps of some of the later designs of locomotive prior to the Standard types introduced by British Railways.

It is extremely simple to use. Once set, trimmed or regulated, it requires no further setting. Open the water valve and check for water at the overflow. Next open the steam valve slightly by about 90 degrees, wait 10 seconds to allow condensation to take place, then open the steam spindle fully, set the flow and the injector will start. To shut off the injector close the steam valve together with the tender or tank water handle.

Possible causes of injector failure are a) no water in the tank, b) insufficient steam pressure in the boiler, c) blocked or loose injector cones, d) vacuum entering the system, and e) the top clack on the injector is shut (where fitted).

After the injectors have been checked the fire can be built up as the departure time approaches. This should be done by placing coal on both sides of the firegrate alternately until the required depth of fire has been reached. This method is used in order to control the smoke and heat being built up in the firebox. If the fire is built up too fast the locomotive may start blowing off excess steam at the safety valves, thus wasting coal and water and making unnecessary noise that annoys the general public.

TESTING THE SAFETY VALVES

When the fire has been built up the safety valves can be tested. On the firebox backplate there can be found a metal plate indicating the registered working pressure and the number of the boiler.

The first safety valve opens when the steam pressure reaches the red mark on the steam pressure gauge, and the second valve opens at 5psi in excess of the first valve, and closes again at the pressure indicated on the gauge. When the pressure has fallen to 5psi below that mark, the first safety valve will close.

If there is more than 5psi difference in the pressure between that at which the safety valves have been set, i.e. the boiler safe working pressure, and that shown on the pressure gauge, the defect must be notified to the driver.

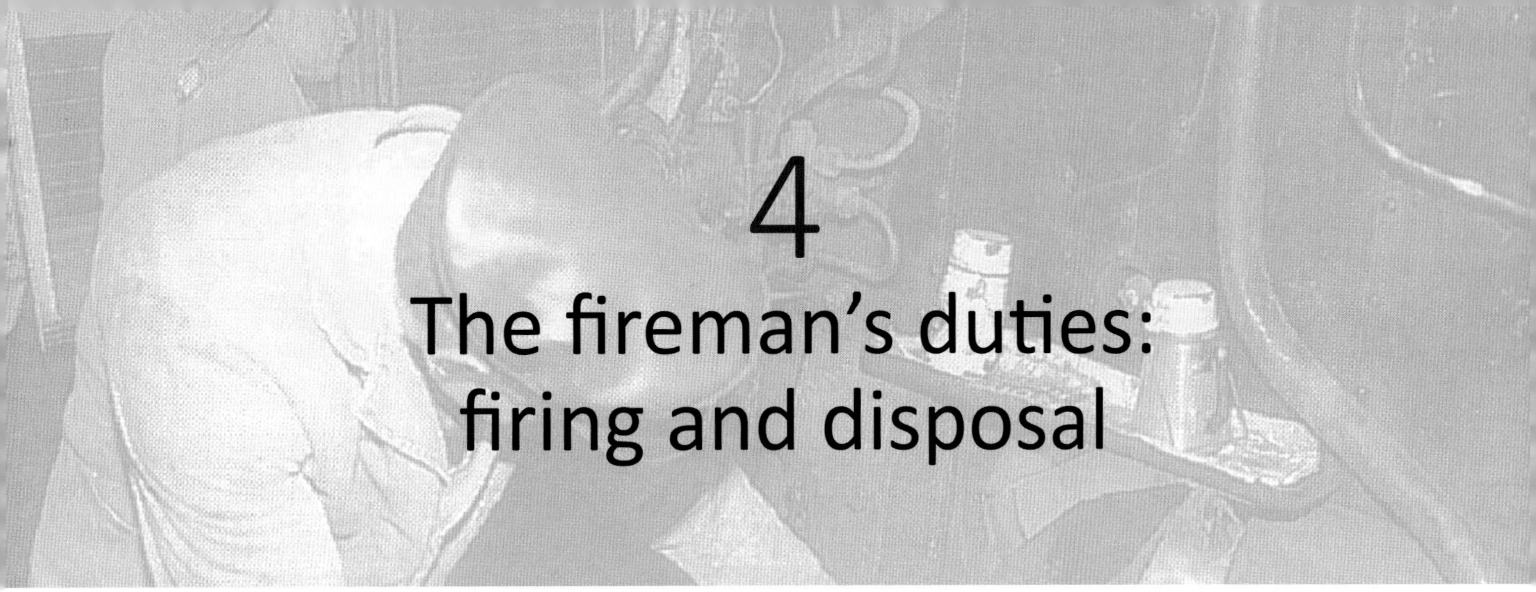

4 The fireman's duties: firing and disposal

Let us now describe a typical day for a fireman on an operating steam railway. This should be regarded as no more than an illustration of what can occur, as a means to explain the fireman's duties and responsibilities; every railway is different, so I can only generalise in order to pass on some of the experiences that will probably be encountered.

THE EFFICIENT FIREMAN

Spending a day as a fireman without the locomotive blowing off at any time during the shift, maintaining the steam pressure in the boiler at all times within about 10psi of the red mark on the pressure gauge, and keeping a suitable water level in the boiler will require a great deal of concentration and a considerable amount of forward planning at all times. However, with experience the fireman will be able to anticipate the actions of his driver and react accordingly.

The fireman must be thinking ahead constantly and planning the journey in order to control the size of the fire. This must be ready for any demand for steam required by the driver, and the boiler must be managed so as to contain sufficient water at all times. About two-thirds full is ideal; this leaves enough space to accept more water if required to stop the engine from wasting steam by blowing off at the safety valves, but without the boiler becoming so full that it primes. A large steam locomotive blowing off at the safety valves will waste 30 gallons of water per minute, and the corresponding amount of coal.

Priming

Priming is caused when the water level in the boiler is so high that, when the regulator is opened, particles of water will be drawn into the regulator valve and along the main internal steam pipe into the cylinders. The symptoms of priming can be detected by a muffled exhaust sound, or signs of water in the exhaust steam at the chimney top. With a locomotive, the drier the steam the more elastic and expansive it will be, and priming, or 'carry over' as it is sometimes called, is dangerous as it can cause the cylinder end covers to blow out due to the hydraulic pressure. Alternatively it can bend the connecting rods for the same reason. The effect of the hot water in the cylinders and steamchest will also act like a steam cleaner, removing all the lubrication from the surface of the bore and causing rapid wear.

An excessive water level in the boiler can also cause water to be drawn from the dome and into the steam supply to the ejector, causing the vacuum to be destroyed and thus applying the vacuum brake and bringing the train to a halt.

Priming can also be caused by the boiler being greasy, brought about by salt deposits in the boiler water, and evidence of this can be seen in white spots deposited on the boiler casing. If this defect is noticed it must be reported so that the boiler can be washed out.

Conversely, of course, the water level should not be allowed to become so low that there is a danger of the fusible plugs melting and causing a failure of the locomotive.

THE JOURNEY

The locomotive for our journey is a Stanier 'Black Five' 4-6-0 and the train is made up of six coaches, for a total weight of about 180 tons. Our journey is approximately 10 miles with three intermediate stops. The gradients are moderate so the run, if we have planned it correctly, will be an easy one.

Depot checks

It is important to arrive at the depot on or before the time specified so as to be ready to take over the locomotive from the preparation crew, who may have already built up the fire. As already described in the previous chapter, the first thing to check is the water level in the boiler by operating the gauge frame. It should be full, or at least two-thirds full.

Attention must now pass to the fire. If it has been spread there must be no blue flame patches that indicate clinker on the grate. The fire should be built up to the required size before departure from the depot, as it is bad practice to do this when in the platform, causing unnecessary smoke in the station. Also, it can be too late for the fire to burn through properly before departure with the train.

Before leaving the depot a check should be made of the

tools, lamps and fire irons to make sure that they are stored in a safe position on the footplate, tender or tank and cannot fall off, and that coal will not drop off and go to waste along the trackside.

The tender must be full of water, otherwise it will have to be replenished before departure from the station. When running tender-first a full tank of water will help the tender to ride better as well as assisting with braking.

The locomotive must carry 'light engine' lamps for the journey from the depot to the train.

Collecting the train

When all the above checks have been made and everything is in order, we can depart light engine to collect the train. If departure from the station is going to be made promptly after arrival, the fireman may build up the fire on the way in readiness. Alternatively, the locomotive may be required to stand for a while warming the carriages with the steam heating on, so again a good fire is essential.

It is now becoming clear how important it is that the fireman gains knowledge of the timetable and the route so that he will know what time and where the demand for steam will be made so as to be ready at the right time and not waste steam.

The fire before departure should be thick enough to maintain steam pressure but not so heavy that the fire cannot breath primary air through the firebed. More firemen experience problems brought about by too thick rather than too thin a fire. However, the fire bed must not be so thin that the blast from the exhaust will blow holes in it and cause cold air to enter the firebox through the grate.

On arrival at the train it is the fireman's duty to hook the locomotive on to the coaches. After the locomotive has buffered up to the train, and with the buffers compressed slightly, the handbrake must be applied, preferably by the fireman. He must then inform the driver that he is going down to couple up.

Coupling up

When the driver acknowledges the fireman's signal, it is safe to go down in between the locomotive and train. First the fireman removes the lamp from the rear of the locomotive and places it in the 'six foot' at the side of the track, pointing towards the locomotive. This is to remind the driver that the fireman is working in between the tender and train. Next, the fireman removes the vacuum pipe or bag from the dummy couplings, first from the locomotive and then the train, leaving them both hanging loose; in this position the vacuum cannot be created by the ejector, so the locomotive will have the steam brake applied, unless it has been clipped up in the off position by the driver.

Screw couplings: The fireman takes hold of the screw coupling with both hands lengthways on the outer sides of the hoop, near the pivot. It is important not to place the fingers near the screw thread but take a firm grip of the outer sides of the shackle; otherwise, should the locomotive move slightly, the fingers could get trapped in the shackle.

With the buffers still compressed the shackle is tightened until the coupling is horizontal, then loosened by about one inch to make later removal easier. If the coupling is left too loose it will cause oscillation between the locomotive and train, while if it is too tight it will be difficult to remove and will also cause wear to the buffers and pins.

Buckeye couplings: As train weights and speeds increased so did the danger of a train telescoping in the event of a collision, with the carriages riding up over one another. The buckeye coupling was designed as a method of keeping a train stable and in one piece; instead of a screw coupling it takes the form

Here the buffers are compressed far too tight, which will cause excess wear on the pins and buffers and will make uncoupling difficult. *Author*

This is still a little too tight - there should be about 1 inch of unused thread. Note that the buffer stocks on the coach and loco are still partly compressed. *Author*

Above: The shackle being tightened: the buffers are slightly compressed so no oscillation will take place between the loco and train. *Author*

Left: When coupling up, the loose ends of the train pipe are kept out of the way clear of the tommy bar while the shackle is being tightened. *Author*

of two large steel hooks, and gives a more comfortable ride for the passengers with less oscillation between the coaches.

Coupling up using buckeye couplings has always been the responsibility of passenger shunters, as the only locomotives to be fitted with them were the LNER 'Pacifies'. However, it is necessary for locomen on an operating private railway to have some knowledge of the buckeye in case of an emergency when a train might need to be split, as follows.

First the train and steam pipes must be uncoupled between the coaches to be detached, and reconnected to the portion to be taken away. Next the electrical connections and the pins securing the flexible compartment ends between the coaches are removed. The rest of the train is secured by means of the hand brake or rail chocks. Vacuum is created on the locomotive, and it is eased back against the braked stock. The ground staff pull the chain to release the pin in the coupling, and a loud click will be heard; the loco can then pull away with the front part of the train. The buckeye coupling can then be lowered by pushing out the securing pin; behind will be seen the hook for the normal screw coupling.

To reconnect buckeye couplings, lift both the couplings and replace the pins, making sure that the pin engage on the flat that prevents it from turning. The pivoting part at the other end of the pin must be down to prevent the pin from sliding out. The buffers must be in the short position with the collars stowed on their pins; when the loco buffers to the train the hooks will be heard to re-engage and the two parts of the train should now be attached. To test the front part, try to pull away from the secured portion; if all is well the train should not be able to move. Reconnect the steam heating pipes, not forgetting the shut-off cocks. As before the train pipe should be reconnected last, the vacuum re created and a vacuum test carried out. In winter months it is also necessary to connect the steam heating pipes. The ends of the steam heating and vacuum pipes cannot be mixed up because they have differently shaped connections; the steam heating pipes have distinctive brass couplings with two hooks. They are kept steamtight by rubber sealing rings on each end, which are retained in a groove and are the same diameter as the rings on the train pipe. The ends are placed together and the hooks closed. This must be done as, unlike the vacuum pipes, they contain steam and with the vibration of the train they could burst apart causing a rapid loss of pressure. Vacuum and steam pipes can be difficult to connect, but the easiest way is to hold one end of the pipe in the left hand then grip the other with the hook. This gives some leverage to get the bottom lugs together and they can then be fastened.

When the steam heating pipes have been connected, the steam valves on the locomotive and the leading vehicle are opened by turning the handles to the horizontal position.

It is always advisable to connect the vacuum pipes last as the train cannot be moved accidentally until this is done. As

The buckeye coupling in the down position, showing the release chain on the right. Note the securing pin and chain below the hook; the collar with the cutaway is at the chain end, and at the other end is the hinged portion that keeps the pin in position. *Author*

The buckeye in the up position. Note the release arm under the hook, which is controlled by the chain; when the chain is pulled the pin is released and the coupling will open when the train is moved away. *Author*

The collar over the buffer stocks. Note the two chains, one carrying the collar, the other attached to the release pin. *Author*

with the steam heating pipes, place both ends of the vacuum pipes together and put in the two spring clips to each of the fixings. If for any reason there is damage to one of the securing clips, it is safe to go with only one as the vacuum in the train pipe will keep the ends together. However, the defect must be reported at the first opportunity.

Like the steam heating pipes, the vacuum pipes are protected against leaks by rubber washers. However, the pipes are not fitted with shut-off valves because Department of Transport regulations insist that the train pipe must be continuous. If taps were fitted they could be left closed accidentally so that even though a vacuum was created on the locomotive, the train would be left with no continuous brake.

The fireman now picks up the headlamp from the ground and sets the appropriate headlamp code on the front of the locomotive. Most private railways have adopted the standard headlamp code that was used by British Railways. Our train is a stopping passenger train - this will require one white lamp on the centre top bracket. After setting the code the fireman places the other lamp on the footplate. Under no circumstances must a lamp be left on the buffer beam in between the engine and train, as if the train becomes divided (although this could not happen with a train fitted with a continuous vacuum brake) it would give false information to the signalman that the train (i.e. apparently a light engine) had passed through his section complete.

If carriage steam heating is being used the fireman turns on the stop cock. In most cases this is on the main steam manifold and is usually a square-shaped valve known as a double seat valve. This type of valve should be opened fully in order to stop steam leaking on to the footplate past the steam spindle. Next the regulator valve is set. A good guide for setting the pressure for the regulator is 1016 for each carriage plus 516 up to a limit of about 5016, although some locomotives have their systems set at 5016.

Preparing the fire

Attention must now pass to the fire. If this was not built up while travelling to the train, it must be made up prior to departure so that it will have time to burn through before the 'right away'. While building up the fire coal is placed on one side of the firebox only and left to burn through. However, it is often said that a fireman should not place coals on the fire when within station limits because it creates excessive smoke; neither must he fire the locomotive or have the firebox doors open while the driver is accelerating the train with the regulator wide open and the engine in full cut off, because this will draw cold air into the firebox and on to the tubeplate.

The way to overcome these problems is the same as building up the fire by firing down each side of the firegrate alternately with the dampers open. Any smoke given off by the fire can be controlled by the use of the blower valve, which

The ends of the vacuum pipes correctly connected with both the pins in place. *Author*

The steam heating pipe and flexible coupling, known as the steam heating 'bag'. Note that the fitting is different from that of the vacuum pipes: there are brass hooks at the end of the bag. At the top is the shut-off cock: up is closed, and down is open. *Author*

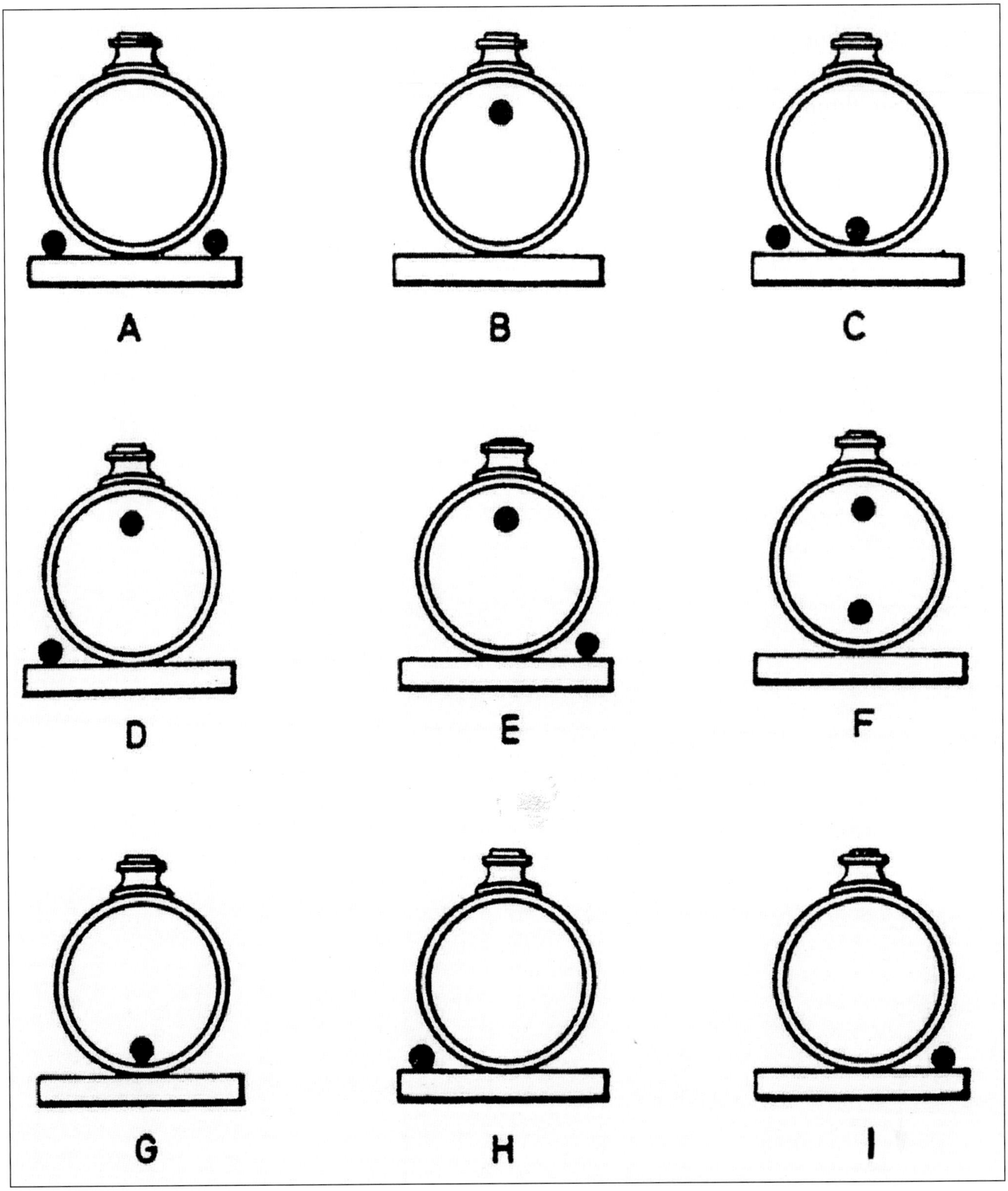

Headlamp codes: A Express passenger train or breakdown train going to a breakdown; B Ordinary stopping passenger train or breakdown train returning to depot; C Parcels, milk, meat or perishables train comprised of coaching stock; D Empty stock or fitted freight with not less than one-third of vehicles fitted with the continuous brake; E Express freight with fewer than one-third of vehicles fitted with the continuous brake; F Through freight or ballast train; G Light engine or engines coupled together; H Through mineral freight or empty wagon train; I Freight train or ballast train for short distances

A locomotive employed on shunting duties must carry one red and one white headlamp at each end of the engine (four in total); the white light must be closest to the running or main line.

Left: The steam heating regulator on the footplate of an LMS locomotive. The pressure is controlled by the wheel on top of the body. The square steam shut-off nut in this example is to the right of the injector steam spindle. Just above is the steam pressure gauge, which indicates the pressure in the system. *Author*

Below: The effect of a badly fitting or burnt baffle plate on the combustion of the fire. Note that the flame path is only passing though the lower smoke tubes.

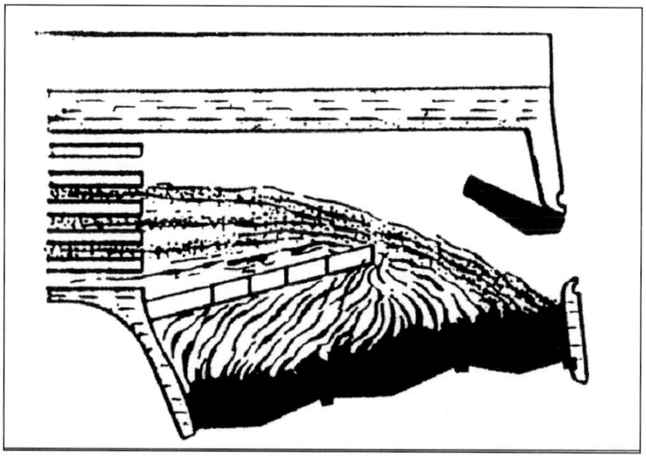

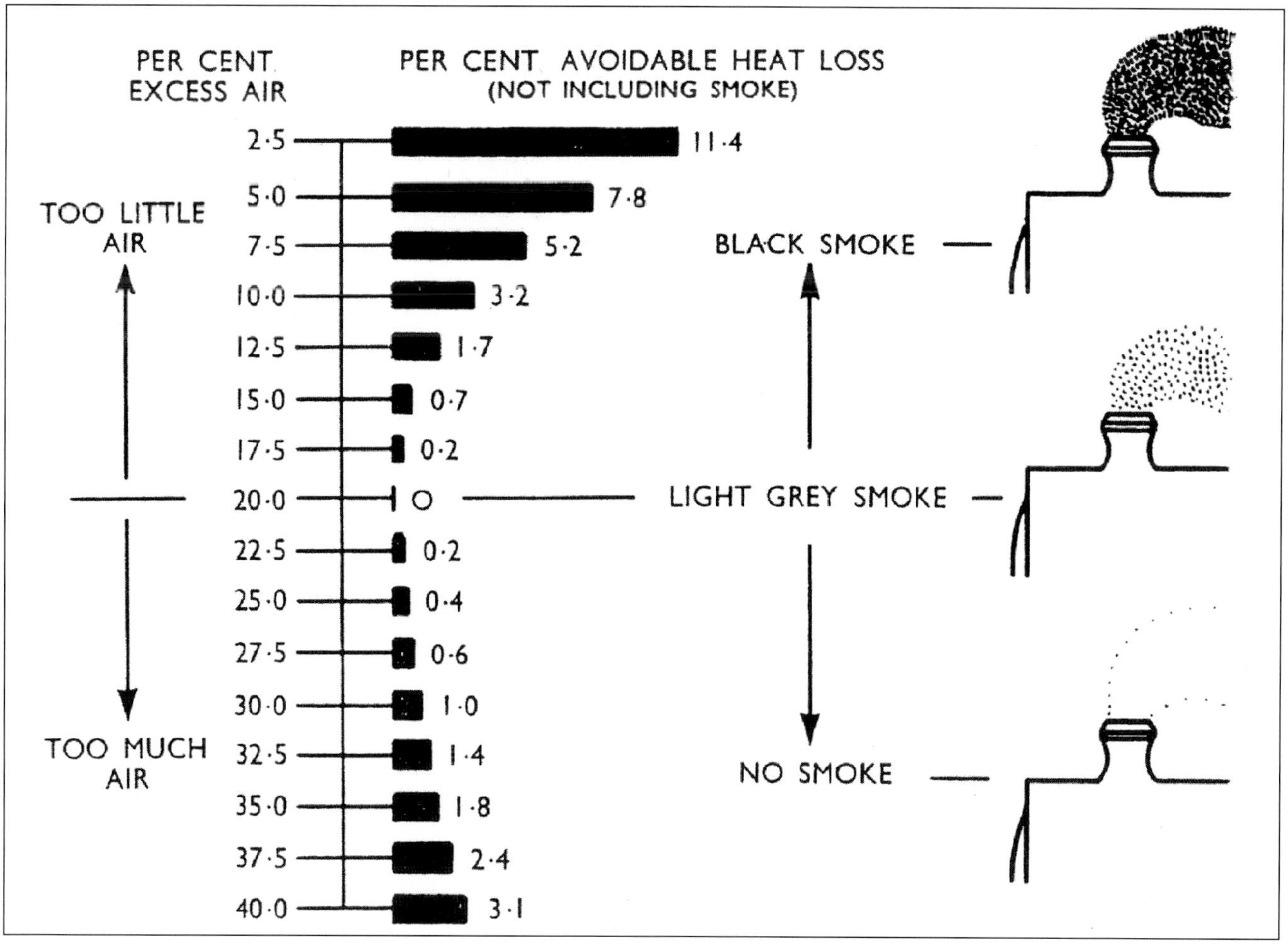

An illustration of the effect of incorrect air being admitted to the grate area of a locomotive. It can be seen that the black smoke given off by the fire due to the carbon not being completely burned wastes the most fuel, while too much air entering the grate allows cold air into the firebox.

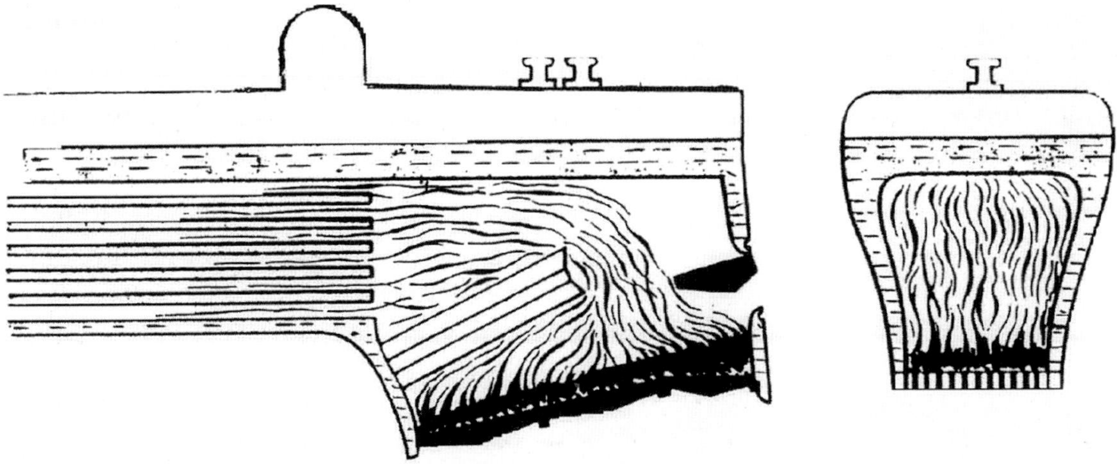

A firebed as it should be, with an even fire all over the grate and no holes or black spots of dead fire. In the combustion space, carbon monoxide is burned and turned into heat.

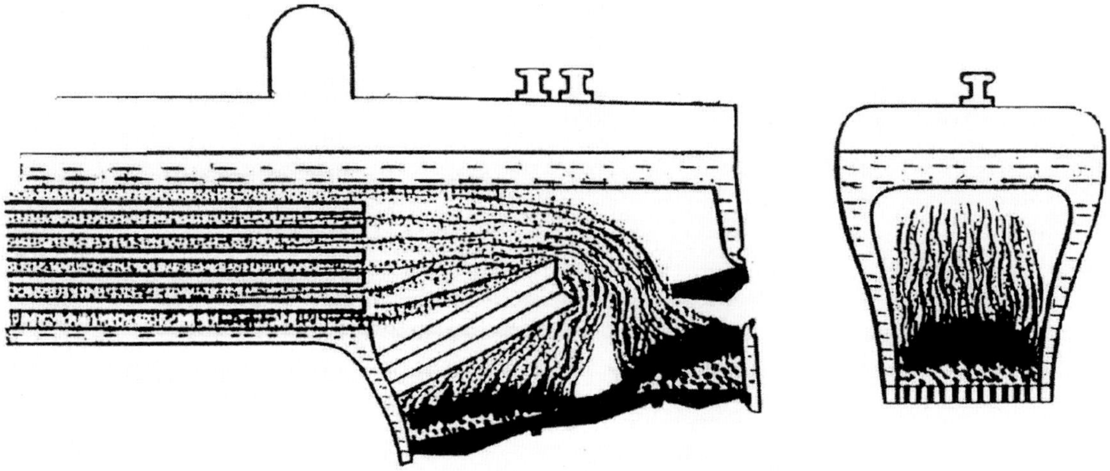

A firebed as it should not be! The fire is heavy, with holes in some places allowing cold air to enter the tubes, and so thick in others that the heat cannot burn the gases in the combustion chamber.

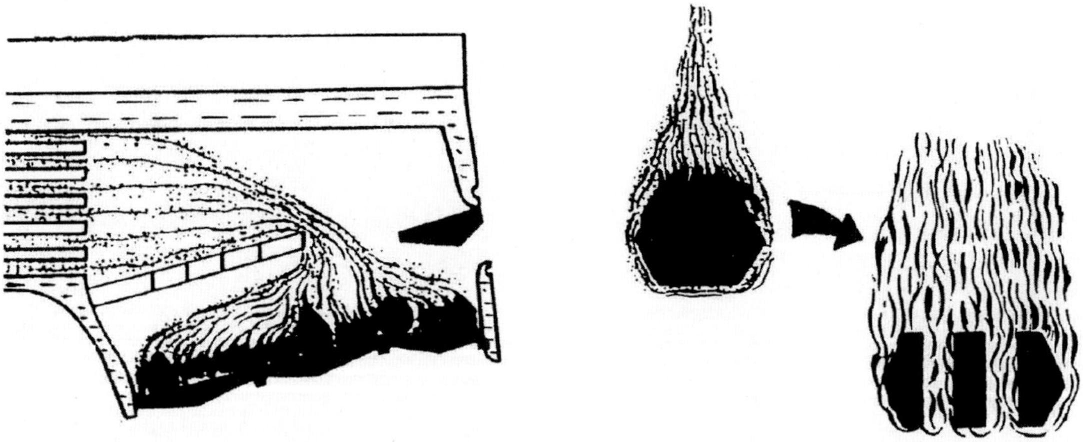

A firebed with large lumps of coal causing black spots where the fire will not make heat to burn the gases. Large lumps should be broken up into pieces no larger than a man's fist. Use the pointed end of the coal pick along the grain of the coal, otherwise the coal will shatter into small pieces. Fist-size pieces burn more efficiently with a better flame size.

Above: Once under way the fireman attends to the fire by placing about eight shovels-full of coal down both sides of the firebox, then closing the firebox doors. *R. G. Fox*

Below: A deliberately staged example of bad firing, causing the production of large quantities of black smoke. *Brian Dobbs*

will make the steam pressure rise quickly. The boiler should not be too full as this will leave room to add more water to control the steam pressure.

The number of ways in which a steam locomotive can be fired is as varied as the engines themselves. The ideal according to the book and accepted science is a thin, bright, level fire with no black spots, but one sufficiently thick to withstand the blast. There should also be no holes that will allow cold air to enter the grate, thereby cooling the fire.

Let us assume that we are approaching the departure time. We should have a nice bright red fire all over the grate area, about 8 inches thick but slightly deeper around the walls of the firebox. With some locomotives such as the 'Black Five' the grate slopes to the front of the firebox, and these locomotives seem to steam better with the fire about the same shape as the firegrate; in other words, a thicker fire under the door sloping to the front of the firebox under the brick arch. Other locomotives require a level fire, while some steam better with a deep fire under the door and a thin fire under the brick arch. Only experience with various locomotives will determine the best methods to be adopted in each case.

Right away!

The departure time has come and the ready-to-start signal has been received from the station staff or guard. When the train starts, both dampers should be open and the firebox doors closed. If the blower valve has been wide open it can be almost closed, but it should never be shut off completely. This is to prevent a blow-back of fire on to the footplate when the driver closes the regulator or if for any reason the smokebox vacuum falls.

It is the duty of the fireman, when not engaged in firing, to look back and watch the train pulling out of the station, as the driver should be looking forward at this time.

Firing

When the train is under way and has achieved the line speed, the fireman must attend to the fire by placing about eight shovelfuls of coal in a similar method as in the station, only firing down both sides, then closing the firebox doors.

The fireman's attention must now pass to the chimney top to note the colour of the exhaust smoke, as this indicates how the locomotive is steaming. It should be light grey, showing that the carbon monoxide being given off by the burning coals that produces the heat is being burned in the firebox.

As carbon forms about 75 per cent of the coal, if there is black smoke it indicates that most of the heat and the fireman's labours are being wasted up the chimney. This is due to insufficient primary air reaching the firebed in order to burn the carbon gases in the combustion chamber above the fire. Black smoke can be controlled by opening the firebox doors slightly for a moment. This increases the supply of secondary air to the fire, which helps to burn away the carbon.

When the black smoke has cleared the doors can be closed until the grey smoke returns. After complete combustion has occurred the smoke will clear at the chimney top and it is time to fire again. By now the firebox and the brick arch should have achieved sufficient heat to overcome the dual cooling effect of the injector placing water in the boiler and coal cooling the fire bed.

It is important before putting water in the boiler to place coals on the fire first, so that heat is being produced as the water goes in; adding water first will cool the boiler, and if coal is placed on the fire afterwards it will increase the cooling effect. It is with this in mind that the fireman must watch the water level in the boiler; if it was full on departure, it should not have dropped to a dangerous level so there should be enough time to place a round of coal on the fire before starting the injector.

With an exhaust steam injector and certain others the flow can be regulated by setting the water handle so that the injector can be left on. This is what is called 'firing against the injector', and is considered the most economic method of working a locomotive.

If after firing there is no smoke at the chimney top, this could indicate that there is a hole in the firebed allowing cold air to enter the firebox through the grate. Further evidence of this can sometimes be indicated by a thunderous roar coming from the firebox, possibly caused by the cold air exploding when it comes into contact with the hot gases in the combustion chamber.

It is possible to look at the firebed against the glare of the fire by placing the blade of the fire shovel on the lip plate of the fire door and adjusting the blade to deflect a shaft of cold air along the firebed. With practice this technique can be used to show the position of the black spot, or where the hole is, or the dead part of the fire. Another indication of the area where the fire is light can be seen at the chimney top; if, for example, no smoke can be seen on the left-hand side of the chimney, the fire will be light on that side of the firebed.

Little or no smoke from the fire necessitates immediate action on the part of the fireman before the firebox cools and the locomotive runs out of steam. One course of action, if there is sufficient water in the boiler to stand it, is to close the injector to arrest the fall in steam pressure. This lack of pressure will cause the level of vacuum to fall, resulting in the brakes starting to be applied automatically.

The lack of steam may be due to a heavy fire, causing a dead or black spot, and as a quick solution it is tempting to try and liven up the fire with the use of fire irons. However, these can cause clinker or ash to be formed rapidly on the firebars, making the trouble worse and starving the fire of air. A better solution is to look at the fire and see the cause of the trouble first; hopefully the problem can be rectified without recourse to fire irons.

If considered necessary, the pricker (a long L-shaped fire iron) may be used to level the fire, the pointed end kept horizontal across the firebed; if the point is vertical it will cause holes to be made. Another fire iron carried on the locomotive is the dart, which can be used to break up clinker from the bars. The pricker now has another use, which is to lift up sheets of clinker from the fire bars to allow primary air to enter the fire bed.

As already mentioned, more trouble has been caused by a heavy fire full of black spots than a thin one. Firing a steam locomotive is not just a case of throwing large amounts of coal

This locomotive is being correctly fired - the carbon is being burned in the firebox. *Brian Dobbs*

into a firebox. A sufficient amount placed correctly will achieve better results than a firebed full of dead fire necessitating the use of fire irons.

It is a bad policy to fire a locomotive just before entering a tunnel, because of the gases given off by the fire in the confined space. Also there is an increased danger of a blow-back caused by the close proximity of the tunnel roof to the chimney top. It is advisable to open the fire doors slightly in a tunnel to provide a little light on the footplate from the glow of the fire.

Assuming that any problems with the fire have been rectified, we now find ourselves approaching the first station stop on our imaginary journey, and we can take stock of the position and leave the fire to burn through. When the driver closes the regulator and the locomotive is coasting, the water level will fall in the boiler gauge glass; the demand for steam causes the water level to rise in the boiler, and when that demand is taken away the level drops. It is hoped that the injector has been able to maintain the water level during the journey, but if necessary any shortfall can be made up while coasting by starting the other injector.

The period of coasting on the approach to the station is also a good time to place a round of about six shovelfuls of coal alternately down each side of the firebox with the blower valve open to control the smoke. This is the opposite to firing out on the line, but the locomotive is coasting so there is less demand for steam. Remember that the action of opening the blower valve will force draught through the grate, maintaining heat in the firebox and helping to raise the steam pressure. It is good policy to keep the fire bed hot and not let it cool down, unless standing, which can cause clinker to form on the firebars.

The stop at this intermediate station may only be for 1 minute, and as soon as the 'right away' has been received from the guard and the train is pulling away, the blower valve can be eased as the exhaust blast will again control the smoke. The boiler water level should by now be sufficient, and the injectors can be shut off to compensate for the demand on the boiler from starting the train. Just as at the start of the journey, the fireman, when not engaged in firing duties, must watch the train away from the platform.

When the smoke has cleared from the chimney it will be time to fire again. The rate of firing will depend on the route. For example, if the line is heavily graded against the train the number of shovelfuls placed on the fire will have to be increased to keep up with the demand, whereas if the line is level or on a slight down grade the rate of firing will be almost nil.

Every line is different, and the successful working of a train depends on teamwork on the footplate between the driver and fireman. If the fireman has little or no experience of the line or locomotive the driver must give sufficient warning as to the route. He must also inform his fireman where the demand for steam will be the greatest, together with any other important details of the journey.

By the second stop the firebox will be hot enough, and an experienced fireman will have control of the boiler and fire.

At the terminus

Let us now assume that we are approaching the terminus of our journey. With less than half a mile to the station the pressure is close to maximum, and the fire is satisfactory, but could possibly take a round of coal. The water level in the boiler is full and the injector is on. This is where planned firing comes into its own: instead of placing more coal on the fire, a better policy would be to close the injector and leave the fire. This will ea use the boiler water level to fall slightly, possibly to half full when the regulator is closed. As the driver is coasting to a halt there will be no call for steam, so the injector can be started, which will prevent the locomotive from blowing off at the terminus, wasting steam and water.

If firing was done so close to the destination, it would mean that the fire would be hot and making steam at the platform, and there would be no room to place water in the boiler to stop the locomotive from blowing off. If fired with a policy of forward planning, the locomotive's dampers can be left open all day, keeping the firebox temperature reasonably constant, with therefore less risk of clinker being formed.

If this is a winter journey, the fireman must turn off the carriage heating control at the stop cock to allow the steam pressure to fall in the system before arrival at the station.

Just as with the departure, it is the fireman's duty to hook off the locomotive from the train, an operation that is carried out in a similar manner to the hooking-on process. When the driver has eased the couplings, the fireman should apply the handbrake and take the spare lamp from the footplate, placing it on the ground in the six foot level with the buffers before requesting and receiving the driver's permission to go in between the locomotive and train.

The vacuum pipes or bags are the first to be split, and the ends left hanging. Next, if steam heating has been used, both steam valves on the locomotive and leading vehicle are closed. When splitting these pipes, take care in case any hot water escapes. The loose ends must be fastened back on the hooks to prevent damage.

Next slacken the shackle on the coupling by using the tommy bar, and remove the hoop from the drawbar hook, taking care again to keep the fingers clear of the screw thread. Place the hoop over the locomotive drawbar hook and secure the end by placing the tommy bar through the hoop. Finally, replace the vacuum pipes on the dummies, carriage first and locomotive last. Place the lamp from the ground on the rear centre bracket and move the front headlamp to the centre bottom bracket to indicate a light engine. There was a rule on British Railways that after sunset or during fog or falling snow all lamps must be lit.

Let us assume that, having travelled boiler-first to our destination, we are going to return tender-first. This will cause any coal dust to blow back on to the footplate, so while running round the train it will be necessary to wash down the coal in the tender and remove any debris from the footplate. This can be done with the slacking pipe, which can be operated when the exhaust steam injector is working and will lay the coal dust and prevent it from blowing into the eyes of the footplate crew. Experience will prove that it is best to wash the coal when tra veiling boiler-first; it is advisable not to do so when the train is standing in the station as any over-spray may splash the bystanders that naturally congregate around the locomotive.

The method of coupling up to the train and firing the locomotive on the return trip is the same as during the outward journey, although it is recommended that the back damper be kept closed to lessen the risk of a blow-back of fire on to the footplate.

DISPOSAL

As this is only an illustration of what can be expected in a day on the railway as a fireman, it is assumed that the locomotive will have made several trips, possibly covering about 100 miles or so.

On the last run of the day attention must be given to the size of the fire so as not to arrive on the shed with a box full of fire to be removed. There must, however, be enough fire to be able to shunt the stock if necessary. There must also be sufficient steam to use the injectors to fill the boiler with water after the locomotive has been stabled.

On arrival at the destination, once the train has been left in the sidings or station and the locomotive has arrived on the shed, it must be disposed in the following manner.

Arrival at the depot

When the locomotive has been secured over the disposal pit at the depot the dampers can be closed and the fire left to cool down. If the locomotive is not required to be used the following day, the clinker and ash can be broken up and lifted from the firebars, then left in the firebox to burn away. However, the ashpan and smokebox will require cleaning before the locomotive is stabled.

If the locomotive is required the next day, the fire must be cleaned in the following manner. The damper doors are closed to prevent the firebox from cooling down too fast and the ash from blowing out of the ashpan all over the motion of the locomotive. The clinker must be broken up inside the firebox by the use of the dart, then the pricker is used to lift the slabs from the firebars. Large lumps of clinker can be lifted out of the fire door on the paddle. Remember that when working in the firebox the shafts of the fire irons will soon become hot. Hands can be protected from the heat by the use of a cotton rag; nylon must not be used because if it comes in contact with the heat it will melt.

To clean the fire start by sliding the paddle across the top, which will remove any good fire on to the paddle; this can then be placed to one side of the firebox. Then use the dart and pricker to place the slabs of clinker on the opposite side, ready to be removed with the paddle. When the clinker is removed, the good fire can be placed under the fire door. The other side of the firebox can now be cleaned in the same way.

Cleaning the rocking grate

Some locomotives are fitted with a rocker-bar grate or drop grate to facilitate the disposal of the fire at the depot. The early type was made up from a section of the grate that was hinged and lowered to enable the clinker to be pushed out through the space into the ashpan. With this early type, the fire can be cleaned by placing an amount of good fire in one corner of the grate and removing the clinker through the space.

The British Railways Standard locomotives and some 'Black Fives' are fitted with a rocking grate that enables the whole grate to be tilted in two separate sections, operated by levers from the footplate (see also Chapter 3). The rocking levers are fitted with a two-position locking clamp; with the locking clamp in the first position the grate can be partly tilted to break up the clinker while the locomotive is moving. When both locking clamps are opened the grate can be fully tilted to enable the contents to be broken up and deposited into the hopper ashpan. Care must be taken to ensure that the locking clamps are in the closed position when the engine is in motion. On BR Standard types, and later LMS locos fitted with rocker bars, it will be necessary to empty the ashpan before cleaning the fire; soak the ash in the ashpan then empty the hopper, closing the doors afterwards. This is done in case the hopper gets so full that the doors cannot be opened later. The ashpan must be closed before the fire is cleaned so that the small particles do not blow all over the locomotive.

Take the paddle and, in a similar manner to the early type of rocking grate, push an amount of good fire under the brick arch in the forward section of the firebox. Then release both sections of the left-hand clamp to allow the back part of the grate to be opened, depositing the clinker and ash into the ashpan. Finish the operation by locking up the grate with both clamps. The good fire can now be placed under the fire door with the paddle, and the process repeated with the forward section of the grate.

Because of the width of the grate, the rocking firebars fitted to the wide fireboxes on BR Standard locomotives tilt independently along each side of the firebox. The method of moving the good fire when cleaning the fire is similar to the early type of fixed grate.

Cleaning the ashpan

The ashpan must next be cleared of all the ash and fire collected during the day, and any residue from cleaning the fire. Before any attempt is made to clean out the ashpan the contents must be soaked with water so that when the ash is raked out it will not blow all over the locomotive. With a non-hopper ashpan both damper doors will have to be opened so that the hose pipe can be inserted close to the top of the ash. Care must be taken not to spray water on to the hot firebars, otherwise damage may occur to their ends. Leave the water running until it can be seen running out of the bottom of the ashpan, then the ash can be removed with the long rake into the disposal pit; careful raking will ensure that the ash will not blow about.

With a hopper ashpan, soak the contents then open the hopper doors using the bar. First remove the latch to release the doors, then rock the doors several times with the bar to shake any residue into the disposal pit. When the ashpan is emptied the doors must be re-locked.

The ashpans fitted to the wide fireboxes have flap doors directly under the grate at the sides, and these must be cleared with a short rake or flushed out with water.

Cleaning the smokebox

Last to receive attention is the smokebox, which will need to be cleaned of ash. Opening the smokebox door has already been described in Chapter 3. Using a long handled brush, draw as much ash as possible to the mouth of the smokebox, where it can be dampened with water before being shovelled out; if it is not dampened, the ash is so light that it will blow all over the locomotive. It is important not to spray water near the tube ends. When all the ash has been removed the lip of the smokebox must be swept clear of ash and any residue swept from the front of the locomotive frames.

Stabling the locomotive

When the locomotive is safely inside the shed, the handbrake must be firmly applied and the locomotive wheels chocked. A steel cover such as a dustbin lid placed over the chimney will retain the heat and allow the locomotive to cool down slowly; less damage will be caused to the boiler if it is left to cool down over a long period of time.

The boiler can now be filled with water, possibly until it is so full that the water level will not come down the gauge glass when the test cock is opened or the injector stops working. When it is full the injector is shut and the gauge frame closed off; leave the test cock also shut, so that if any water passes the shut-off it will build up in the gauge glass and not escape.

If the steam heating has been used, the stop cock on the manifold must be closed off. With certain locomotives, especially those with faceplate-mounted injectors, it will also be necessary to close both injector top clacks on the injectors.

Before leaving the locomotive, make a final check that the dampers and firebox doors are closed and the blower valve shut off.

It is traditional that the driver and fireman exchange thanks after a good day, and there is no better way to do this than over a pint of good beer - as long as there is no need to drive home!

Learning by experience

One of the best ways I can illustrate how a fireman gains practical experience on the footplate is to recount two incidents from my early days on the footplate.

In the first, when I fired on a passenger train for the first time, my driver John and I were booked to cover a special train. On our arrival at Bury shed John looked at the engine duty board to find that our loco was on road three; to my delight it was a Horwich 'Crab', No 42719, standing tender-first up the shed. I went for the oil, and on my return we went about our duty and soon had the engine ready for the journey.

My mate told me to release the hand brake and we made our way to the signal box. On arrival John told me to inform the signalman where we were going and for what purpose, which was to assist a passenger special to Accrington from Ramsbottom. Soon the signal cleared, my mate opened the regulator and we made our way light engine up the line to Ramsbottom. It was the first time I had worked on this type of locomotive, although we had five of the class stationed at Bury. We used them as a general-purpose locomotive for fast freights, parcel trains or the occasional weekend special to the coast.

On the way to the station I built up a good fire, or so I thought. I fired under the door, leaving the front end light as with the 'Derby Four' earlier in the day. Soon we arrived at Ramsbottom and were placed in the sidings at the north end of the station to wait for the appearance of the special. The boiler was full and the locomotive was blowing off occasionally, so in my inexperience I put the damper down to let the fire cool a little. After a diesel railcar arrived and departed on its way to Bacup, the signals cleared for the special, which arrived, hauled by a Caprotti 'Standard Five'. The train stopped in the platform, an engine's length behind the home signal just before the level crossing. John told me to lift the damper and to place some coal under the brick arch and under the door. Then the ground signal cleared and we reversed back on to the train.

It is the duty of the fireman of the assisting locomotive to hook on to the train engine. As the buffers came in contact I dived in between our tender and the train loco. As soon as the vacuum was destroyed I took off both the vacuum pipes, threw our shackle over the drawbar of the train engine and screwed it tight, then finished by connecting the train pipe. The other fireman removed his headlamp, although I did not actually see him. The last thing to do was to put the headlamp from our tender on the front buffer of our loco.

I had steam pressure on the red mark. It is the duty of the leading locomotive to create and maintain the vacuum level, and it also has control of the brake. My mate opened the large and small ejectors and the vacuum clock soon indicated 21 inches; the large ejector was then closed and the level remained steady.

The signal was clear, so John and the train loco driver exchanged whistles, then opened their regulators. There was the sound of escaping steam from both chimneys, and the 12-coach train moved slowly forward under the dual effect of both locomotives. As soon as the train was under way John wound the reverser close to mid-gear, fully opened the regulator to second valve, then adjusted the valve travel. The locomotive then started the front-end roll or side-to-side motion that gave the 'Crabs' their nickname. Just 20 or so feet at the rear of our tender the 'Standard Five' could be heard working hard. It felt great to be a fireman of a passenger train!

Passing the coal yard I caught sight of the distant signal for Stubbings Junction, which was showing clear; here the line diverged to the right to Bacup and to the left to Accrington. The gradient along the latter line included a section known as Baxenden bank, which had a ruling incline of 1 in 74 for 5 miles from Stubbings Junction as far as Baxenden, where the line started to fall for the last 2 miles into Accrington at an incredible 1 in 27.

We had just passed Stubbings Junction signal box when I noted with horror that the water level in the boiler had fallen to just below half, and this with the regulator open. It was incredible that in such a short distance the water could fall so much - we had travelled no more than a mile. I suppose it was possible that I could have misread the water gauge before getting off the locomotive to couple on.

This was serious, and in a state of panic I put on the injector to maintain the water level. The cooling effect of the water in the boiler brought down the steam pressure, so in a further state of alarm I attended to the fire, which compounded the mistake as my actions also had a cooling effect on the fire. I used the shovel blade to deflect cold air across the fire to see into the firebox. Black coal could be seen on the firebed causing cold areas in the grate. The injector should have been shut off to allow the firebox to warm up, but in my panic I took the dart from the tender to stir up the firebed. This was another mistake - the use of the dart caused clinker to form, again cooling the fire.

John's first indication that I was in serious trouble was when the vacuum level started to fall, caused by the low steam pressure. He came over to my side to help, but things were bad. The vacuum brake was coming on, so action was needed. My mate returned to his side and indicated to the driver of the train engine that he should create vacuum; this done he closed our ejector, saving our steam and releasing the vacuum brake. Looking at the boiler he told me to shut off the injector for a moment, which did allow the steam to raise a little.

By now we were approaching Helmshore station and our pressure was down to about 120psi, 60psi below the maximum, but it increased to 130psi with the injector shut off. Soon we passed Grain Road level crossing, and to my relief the steam was continuing to make a little progress, but the water was dropping in the glass. Things were desperate indeed.

John tried to assist me, but despite his valiant attempts he made little difference to our predicament. The boiler water level now was almost in the bottom nut, nearly empty, so the injector would have to be re-started soon, but the steam had raised to 140psi. On John's instruction I placed a round of coal on the fire and a few moments later re-started the injector. The only thing that could be done was to mortgage our requirement for steam against the steam level and water to keep us going.

We had to get more water in the boiler before the summit and the following 1 in 27 descent. Running downhill the water would move to the front of the boiler, and with our low water level it could leave the firebox crown sheet uncovered, therefore causing the fusible plugs to blow and failing the loco. Nearing the summit our speed was dropping, and the 'Standard Five' could be heard to open up a little more to compensate for our poor performance, doing more than his share of the work.

Soon the summit appeared, and as we passed Shoe Mill signal box the water flowed to the smokebox end of the boiler and was completely out of sight in the gauge glass. As we were now coasting there was no requirement for steam, so both injectors were started, and soon there was water showing in the bottom of the gauge glass. The blower valve was fully opened to help raise the steam pressure. We coasted through the tunnel to a stand at the station at Accrington with about half a glass of water, so I left one injector on.

Just as it was my duty to hook on, it was also due to me to hook off. Meanwhile John said that he would look after the boiler. I was in between attending to the couplings when the other fireman appeared. He was heavy built, about 30 years old, and covered in perspiration from the hard last part of the journey. He looked at my new overalls, possibly indicating my lack of experience, and, placing his headlamp on the bracket, he said, 'Not done much, have you, lad?'

'No,' I replied sheepishly. 'Never mind, you'll learn.'

With this he went about his duties and I returned to our engine. The signal cleared and we went round the triangle to turn the loco and return light engine to Bury. The fire was now blue, indicating heavy deposits of clinker and ash on the firebars. Running light we soon made the top of the bank at Baxenden. Most of the fire had burned away and as we coasted down to Stubbings John said to run the pricker through the grate to try to clear the firebars of dross. This was hot work, but when I finished the fire looked clean and the pricker was replaced on the tender. A light firing brought the steam pressure back to the red mark. While I was filling the boiler I washed down the footplate ready for our relief.

'Crab' 2-6-0 No 42700 passes Bradley Fold Junction, where the line starts to descend the gradient into Bolton. *Eric Bentley*

The second incident took place the following week. I had been on the shed cleaning locomotives, which was not a bad way to regain my confidence. On Monday I met some Blackpool men when they came to collect their loco to work a passenger train from Rochdale to Blackpool. I was fed up with cleaning and decided to assist them to prepare their engine. Later, having a brew with them before they left to go light engine to Rochdale, I related the disastrous events of Friday night with my first ever passenger run.

Being out firing I didn't see them again until Thursday. I was back on shed cleaning that day as no footplate work was available. I soaked the motion of the loco I was working on, leaving it to dissolve the grime while I went to the stores for the lamps for the Blackpool train. Harry, the foreman, saw me filling and cleaning the lamps and inquired what locomotive they were for. Somewhat sheepishly I replied that they were for the Blackpool loco. He just looked and said, 'OK, so long as you are working', then went away.

Soon the voices of the locomen could be heard approaching. Picking up the lamps from the bench I accompanied them to their loco. They placed their mess bags in the locker on the tender then went about their duties making the loco ready. The loco was over the pit and I was expecting the driver to ask me to oil the inside motion as I did on Monday, but the fireman was happily oiling the loco. The driver checked the water level in the boiler and the steam pressure, then looked in the firebox.

'Will you put the lamps on the brackets?' he requested.

I duly put up the light engine headcode (one lamp on the bottom centre bracket front and rear). Unnoticed, Harry was watching me from the other road, and when I returned to the footplate he appeared at the steps of the loco.

'Are you helping these lads?' he shouted up.

'Yes,' I replied.

He walked away a few steps then turned and suggested that I ask the driver if they would take me with them as far as Bolton to learn a little more about the job. I must have looked puzzled. Had the disastrous events of last week come to his attention or was it just a convenient way to keep me occupied? I looked at the driver.

'Don't ask me,' he said. 'It's my day to fire- you'd better ask my mate.'

The fireman had returned to the footplate and had overheard the conversation.

'No problem,' was his reply.

'All right,' said Harry. 'Don't forget how Bolton is spelt- B-O-L-T-0-N - because you're finishing the day helping the disposal crew.'

Under the driver's instruction I helped build up the fire with coal all over the firebox and the damper down; this was then left to burn through slowly. Because I had gone for the oil and filled the lamps before their arrival, we had time to fill the tea cans. Soon we had the loco ready and we left tender-first for Rochdale, where we collected the coaching stock. The driver stood at the side while I hooked on and placed the head code for a stopping passenger train (one lamp on the top centre bracket just in front of the chimney).

On the way the driver kept a low water level in the boiler. I asked many times whether I should put on the injector, but he said, 'No, let's keep the boiler low on water.'

'Why?' I asked.

'You'll see,' was his reply.

As we stood in the platform waiting for the 'right away', with no need for steam and a large fire, the steam pressure was raising fast and was soon on the red mark.

'OK,' said the driver, 'put the injector on to control the boiler.'

The reason for the low water was now clear, and with the feed on the steam pressure fell a little to about 200psi. Just before we started he told me to open the dampers and start the dual live and exhaust steam injector, as the fire would now make steam quickly. Soon we got the 'right away' from the guard and left Rochdale on time.

'Close the live steam injector - my mate needs the steam - and watch the train out of the platform.'

The fireman drove the engine well and soon we were at our first stop at Castleton. When the fireman shut off, the driver said to place a light round of coal on the fire, not forgetting behind the doors. The heat was intense with the draw of six exhaust beats to each revolution of the wheel. I placed six shovelfuls on the fire as we came to a stand and closed the doors, then adjusted them to control the smoke.

'OK,' said the driver, 'check the boiler while we are standing.'

'Half full,' was my reply.

'While we're standing brush the loose coal from the footplate.'

A blast on the whistle and we were away again. I watched the train out of the platform, and when we were clear of the station I washed down to lay the dust. It was fantastic having someone to guide me in the duties of a fireman.

We stopped at Heywood then Broadfield, and soon we were going down the bank before Bury. I was now getting into a routine unprompted: I put plenty of coal on all round the box, not forgetting behind the doors.

'Right, we're coasting now,' said my instructor. 'Put the live steam injector on while there's not a great need for steam.'

We stood at Bury Knowsley Street station and the loco blew off. A blast on the guard's whistle, replied by the fireman on the loco whistle, and we were off up the bank to Bolton with a stop at Radcliffe Black Lane and Bradley Fold, then down the bank into Bolton Trinity Street station. This was a good time to clean up ready for the driver to take over.

When we stopped at the station I collected my brew can from the warming plate and bid goodbye to my friends. I then left the loco to travel as passenger back to Bury to finish the rest of my shift on the disposal pit. I never did know the crew's names, but learned more in that short run than I had managed previously.

There's no better way to learn than by practical experience. While I worked on the railway we had to read the Rule Book, which meant nothing in isolation. However, some years later, after reading L. T. C. Rolt's book on railway accidents, *Red For Danger,* every rule made sense. I hope these experiences will do the same.

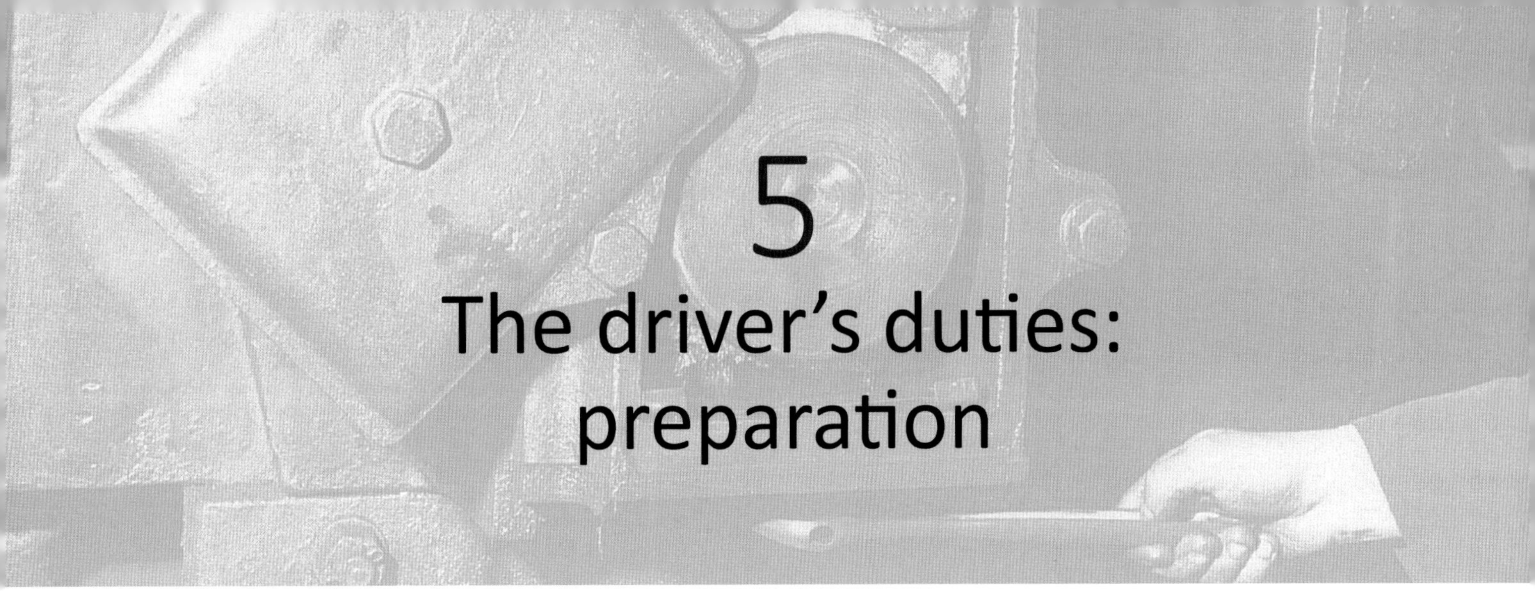

5 The driver's duties: preparation

THE EFFICIENT DRIVER

The duty of a driver is to drive the locomotive or train in a safe and correct manner, and to maintain the timetable without taking any risks with safety. He must obey the rules of the railway and the signals, and must drive in a safe and economic manner so as not to be wasteful of fuel or cause damage to the locomotive or train by misuse or hard driving. He must also not take over the locomotive if he is under the influence of alcohol, or drugs, that will effect his judgment as a driver. It should always be a driver's intention to improve his knowledge of locomotives and of the railway in general. The basic training that he will receive on being promoted from a fireman will stand him in good stead to carry out his duties if nothing out of the ordinary happens, but a driver who has gained extra training will be in a better position to drive the engine economically and react in the case of a problem with the locomotive or train, or to any emergency that may develop during his day.

The driver is at all times responsible for the actions of his fireman and must supervise and advise him about his duties, whether in the preparation of the locomotive or out on the road. During their shift together he must look after his fireman and develop the special teamwork that is required for successful locomotive working. He must inform an inexperienced fireman of the requirements for steam and water along the route, and by supervision ensure that he does not waste fuel and water by allowing the boiler to blow off at the safety valves. He must also check that the fireman is keeping a sufficient and safe water level in the boiler.

As in the previous chapter I will describe a typical driver's day, dealing first with the preparation of the locomotive. However, I will deal in greater detail with the different locomotives likely to be encountered, because the various designs require different methods and actions on the driver's part, such as the difference between the steam brake and vacuum brake of the locomotive and train.

To some extent, just as the various types of locomotive require different firing methods to achieve the best results, so there are different locomotive driving techniques. However, the various methods of firing will not necessarily affect the safety of the train and the passengers, which is of greater concern to the driver.

This chapter describes the preparation of locomotives associated with the former 'Big Four' companies and British Railways Standard designs. These, together with their design derivatives, will cover a considerable amount of the locomotives in use at present on the operating preserved railways in the British Isles. For example, they share basic similarities in the motion, the Walschaerts valve gear fitted to former LMS engines being the same design as on LNER and Southern locomotives and BR Standards, with only a few slight modifications, the principal differences being in the slide bar and expansion link arrangements.

VISUAL EXAMINATION OF THE LOCOMOTIVE

On arrival at the depot the driver must first sign in, then locate the correct locomotive and begin to prepare it for the day's work. On arrival at the locomotive he must carry out an inspection to ascertain its suitability for use.

As he climbs on to the footplate the driver, like the fireman, will inspect the boiler for sufficient water and steam; it is quite possible that the fireman will have already done so, but it becomes second nature to locomen that when a footplate is entered the first action is to check the water level in the boiler.

Before any work is carried out on the locomotive it is important that the handbrake is applied and the wheels are chocked so that it cannot be moved. 'NOT TO BE MOVED' boards should be placed on the cab sides or the lamp brackets protruding outwards, warning others that staff are working about the locomotive.

While the fireman goes to the stores to collect the oil for the engine, the driver can start his visual examination. He must first check the springs for broken leaves. This emphasises the importance of a clean locomotive - it would be easy to miss seeing a defect in places such as springs that may be covered with a layer of grime. However, a broken top or main

An assortment of oiling cans. At the back are two sizes of oil bottles used to store the oil on the footplate. At the front can be seen two types of oil feeder cans; that on the left has a short spout, while the one on the right has a long spout ideal for difficult places. *Beesley Products*

A flat oil can; this is ideal for inside big ends as it will fit between the crank webs. *Beesley Products*

leaf is easy to see, and will mean that the locomotive is a failure and cannot be used until the spring has been replaced. If there is only one broken leaf in the main body of the spring the fitter on duty must be informed so that he can decide if the locomotive is safe to go into traffic or not, or whether the spring can be replaced before it enters service.

As the driver works round oiling the engine he will look for possible defects such as missing cotter pins, loose nuts, broken rivets, and split or frayed vacuum, steam heating and brake pipes.

When the locomotive is in steam the various shut-off cocks will need to be opened on the footplate, such as the sanders, whistle, steam gauge, steam brake and atomisers. Some of these are duplicated; for example, the atomiser shut-off cock on LNER locomotives also shuts off the whistle. There may also be shut-off cocks on the outside of the footplate, such as the main steam shut-off on a Southern Railway 'Pacific'. On LMS locomotives the atomiser shut-offs have the tell-tale escape of steam from the centre when the valve is closed. There are also the oil-warming cocks fitted close to the body of the lubricator. If some item of equipment will not operate, first look for a shut-off cock by tracing the pipework back to the source of the steam; if there is a valve in the system it will be readily seen.

LUBRICATING

By now the fireman will have returned from collecting the oil and, if the locomotive is in steam, will have placed the can on the warming plate over the fire door. He will then start his own duties in the preparation of the locomotive. The oil is left to warm through so that the heat will lessen its viscosity and it will flow more easily into the oil cups; this is particularly appreciated when filling difficult places like the driving wheel axleboxes on Great Western locomotives such as those of the 'Manor' class. Due to their design the filling point is small and mounted on the side of the axlebox. There are two fillers, one at the top and one at the bottom between the axlebox and the hub of the wheel; filling them is best likened to filling the sump of a car engine through the dip-stick hole!

A useful technique is to use two oil cans, one warming on the manifold while the other is in use.

Types of lubricating oil

The reason for oiling a locomotive is to provide a film of lubricant between metal surfaces to prevent the friction that would otherwise cause wear to the bearing faces.

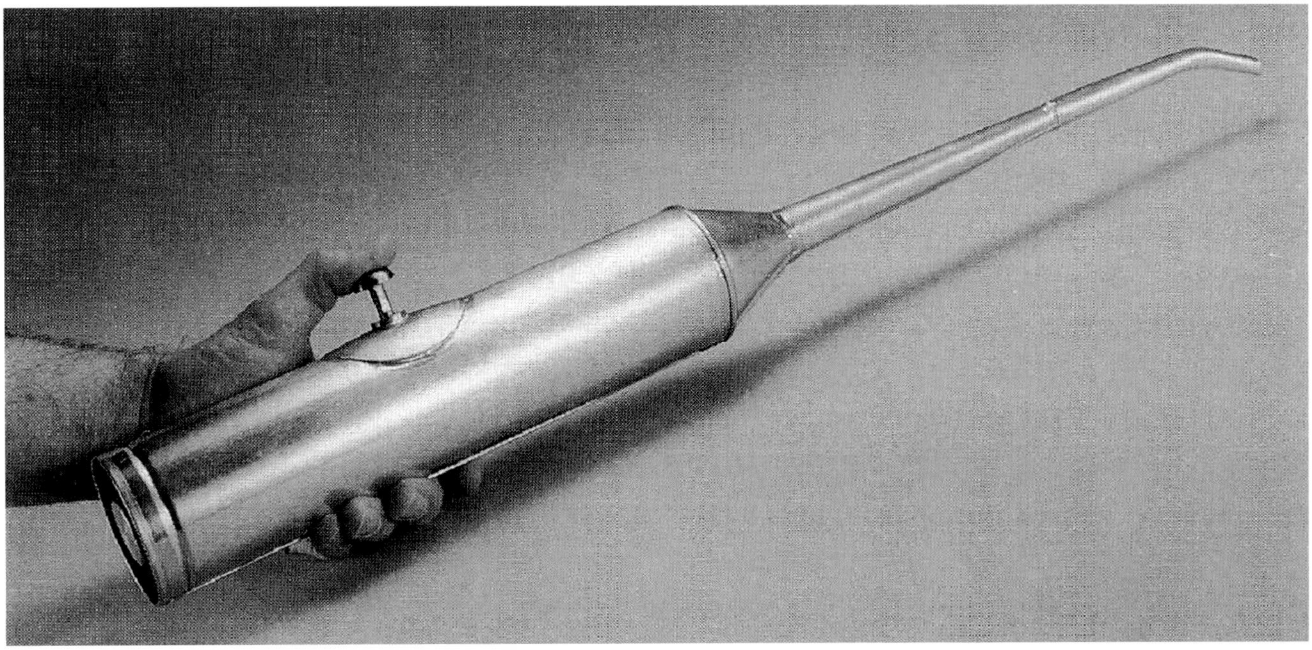

A barrel-type oil feeder with valve. This can does not pump oil and must not be confused with the syphon pump. *Beesley Products*

The work carried out by the manufacturers of lubricating oil has brought about many changes in the way a steam locomotive is lubricated.

The oil used for the oil boxes or cups on the frames is mineral-based, selected because it will syphon from the cups. Special additives mean that its consistency is thicker than in British Railways days, and when it is poured from the can to the oil cup it leaves behind a slightly sticky thread of oil that remains on the bearing surfaces.

There used to be two grades of cylinder oil in common use on BR, for superheated and saturated steam locomotives; the heavier grade was used on the superheated locomotives because of the greater heat. Today only one oil is required for modern cylinder lubricators, whether hydrostatic or mechanical, and it will cope with both superheated and saturated engines. The developments mean that the new lubricants carry more detergent properties and do not carbon up the valves like the old types; they also provide a better film of lubricant on the bearing surfaces.

The oil for the lubrication of the steam brake is the same as used for cylinder lubrication, and is consequently known as cylinder oil. Again, the heavy viscosity is to contend with the heat of the steam in the cylinders. It is usual to fill the mechanical lubricator for the cylinders and the steam brake first so that the oil can warm through thoroughly and thus flow more easily.

Oil feeds to moving parts

Besides lubrication by the hydrostatic or mechanical lubricator, there are three other types of lubrication, for all of which engine oil is used; these are lubrication of sliding, reciprocating and rotating parts.

For sliding parts, such as slide bars and crossheads, the lubricant is usually fed from above, with the oil contained in a cup or oil pot and fed by a drip feed. The oil is syphoned out of the cup by various types of worsted trimmings, as shown in the accompanying diagram. As the sliding part passes the feed hole, a drop of oil is fed to the bearing surface by the trimming. There are three types of plug trimmings:

Plug: The plug fits in the oil hole and restricts the flow of oil. Plugs are found in parts such as the pivot bearings of the expansion link.

Plug tail: This trimming is a plug with a tail that draws oil from the pot to the lubrication point to provide a steady flow.

Tail: This plug trimming has two tails to provide a greater flow of oil to the lubrication point.

Tail trimmings are usually used in oil cups that are remotely mounted on the main frames of the locomotive, where the oil is fed down pipes to the lubrication point.

On parts such as the expansion link, which does not rotate completely on the trunnions, 'reciprocating' lubrication is employed. The oil is kept in an oil pot usually cast into the mounting bracket and forming a well from which the oil is fed. The pot is filled with a worsted pad rather than a plug, which receives and stores the oil and allows it to pass slowly to the lubrication point. Two types of pad, long and short (the latter sometimes known as a mop trimming) are shown in the accompanying diagram.

Where a part is rotating, in such important places as the big end of the connecting rod or in the eye of the coupling rod, a greater amount of oil is required. The oil for these parts is contained in a large cup carried on the rod itself and sealed with a cork. Inside the cup is a restrictor plug that controls the flow of the oil; some are of the screw type while others are push-in with three flutes cut into the side. As the part revolves the oil is thrown about inside the cup, causing splashes to drip past the plug on to a felt pad below, which keeps a supply of oil on the bearing.

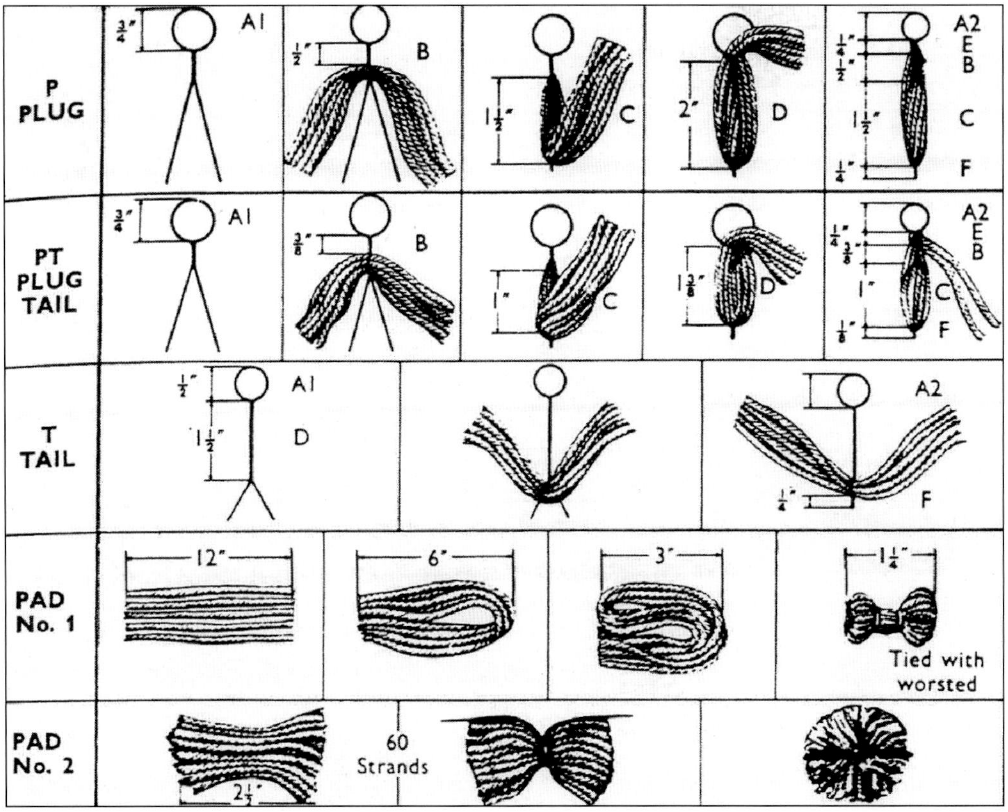

Left: Diagrams of how to make the worsted trimmings used for locomotive lubrication.

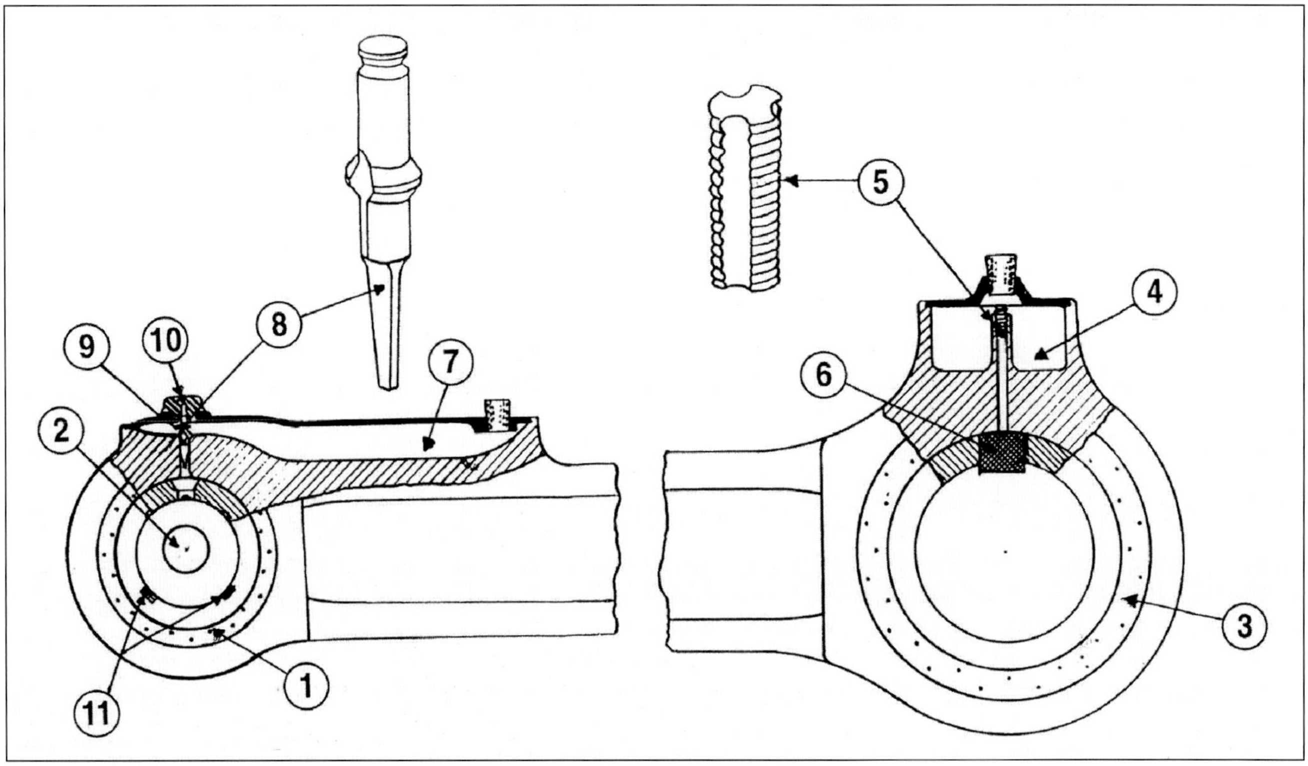

Below: Diagram of the connecting rod of a locomotive, showing two types of restrictor: 1 little end side face with lubrication holes; 2 little end eye; 3 big end side face; 4 big end oil cup; 5 screw-type restrictor; 6 felt pad retaining oil supply for big end bearing; 7 little end oil cup; 8 push-in restrictor; 9 little end oil feed hole; 10 bolt allowing access to restrictor; 11 felt pads retaining oil supply for little end bearing

LUBRICATING AND INSPECTING VALVE GEAR

The oiling and preparation of a steam locomotive must be carried out with care and detailed observation of every square inch of the frames where oil points may be concealed. As already mentioned, the task will be made easier if the locomotive is clean. Always start in the same place, whatever the type of locomotive being oiled. Most drivers start at the outside left-hand trailing coupling rod and work round the engine in a systematic manner, finishing with the outside right-hand trailing coupling rod, before starting on the inside motion.

A view of the Walschaerts valve gear of Flying Scotsman with the locomotive in the correct position for oiling. Note that the big end of the connecting rod is in the top back quarter, making all the oiling points on the motion accessible to the driver's oil can. *Brian Dobbs*

Two-cylinder Walschaerts

When oiling the motion of a two cylinder steam locomotive fitted with Walschaerts valve gear, the task is made easier if the locomotive is set with the motion in the correct position. The big end of the left-hand-side connecting rod should be in the top back quarter, as this will place the oil cups of, for example, the leading crank pins on the coupling rod in a position where they can be reached without re-setting the engine. Similarly, the filler of the oil cups of the little end will be tilting backwards in a position to hold the maximum amount of oil, and the big ends will be easy to fill, being clear of the radius rod.

As already mentioned, make sure that the oil is warm so that it will flow more easily into the lubrication points. Before starting to fill the feeder can, wipe it clean so that it will not spread dirt that might contaminate the oil before it is placed into the lubrication points.

Start with the back coupling rod on the left-hand side of the locomotive. Each filler hole is sealed with a tapered cork; first wipe round the base of the cork with a clean rag to prevent any grit falling into the oiling point when the cork is removed. Screw the cork out of the filler and top up the cup with oil to the base of the thread. Inside this type of lubrication point will be found the restrictor plug; as the oil cup is filled with warm oil it will run over the hole containing the restrictor and the excess will flow down the side of the restrictor flutes, giving the bearing an initial oiling to the worsted pad.

If the oil cup is accidentally overfilled, push a finger into the orifice to the base of the thread; this will remove the surplus lubricant. If it is not removed it will force out the filler cork and the revolving motion will spill out the rest of the contents. Replace the cork, making sure that it is firm and straight in the hole, then wipe any excess oil from around the hole so that it will not attract dirt or grit.

With some locomotives the usual cork filler is replaced by a design that utilises a spring-loaded ball to keep the oil in the cup and the dirt out. This type can be seen on such locomotives as the former LNER 'Pacific' *Flying Scotsman.* To fill this type requires the use of a screwdriver or a 3- or 4-inch nail to press down on and hold open the ball. First wipe round the base of the filler with a clean rag, then press down on the ball to open the orifice; when the cup is full and the pressure on the ball is released, the spring will automatically return the ball to its seat. Any overfilling can be wiped off with a clean rag as before.

While oiling the locomotive, check that all the taper pins that retain the nuts and collars, such as those on the rear crank pin, are in position. Looking at the thick end of the pin, there should be no trace of bright metal showing at the bottom of the pin close to the collar or nut, which would indicate that the pin is loose in the hole.

A check must also be made that all the split pins that hold the various nuts secure are in place and the ends open. Here again the benefits of a clean locomotive are evident, making the detection of faults such as loose pins easier. A loose or broken rivet in the main frame can be detected by a trace of rusty water bleeding out on to the paintwork.

While inspecting the wheels a trace of bright metal around the balance weights can again indicate that they are loose. Broken rivets on the wheels can also be detected by a trace of rust around the ends of the rivets. A check must also be made for a loose Gibson ring on the inside of the driving wheels.

The crank pins, eccentrics, axleboxes and big ends must be inspected for traces of white metal, indicating a hot bearing possibly caused by a lubrication problem.

Working round the locomotive in a systematic method, the next oiling point will be the pivot at the end of the coupling rod by the big end. This is a simple hole in the end of the rod and the best method of lubricating it is with a pressure oil can, preferably with a flexible spout. The reason for the rod being able to pivot at this point is to allow it to flex when the locomotive negotiates a bad joint in the track. Again there is a nut retained by a split pin, which must be checked as the pivot point is lubricated.

Also using the pressure oil can, lubricate the brake pivot pins both on the stretcher and the brake block pivots in the hangers. At the same time, if the locomotive has a tender, a check can be made of the adjustment of the brakes. This can be carried out as the handbrake of a tender locomotive operates only on the tender wheels, so the locomotive brakes can safely be off. The brake blocks should have sufficient metal and be free of contact with the driving wheels in the off position, but still in close proximity to the tyre. As the handbrakes on a tank engine work on the driving wheels, the brakes must be inspected with the locomotive chocked and the handbrake in the off position.

The next point to be oiled is the crank pin that carries the big end of the connecting rod, the coupling rod and the return crank. Between the coupling rod and the return crank on each side of the big end bearing can be found a thrust bearing, which must be lubricated by pouring oil in between the faces. There will be some clearance between all three components, but it should not exceed about a quarter of an inch.

The return crank is retained on the crank pin by one of two methods. The first is with a key-way machined into the face of the crank pin and secured with four nuts; again, as these oiling points are being attended to, the nuts must be checked for tightness. With the other method the return crank is fitted on a square on the end of the crank pin and secured by a cotter bolt. In our systematic progress attention must now pass to

Top to bottom Oiling the return crank of Walschaerts motion. First wipe round the cork to prevent any grit from entering the oil point, remove the cork and fill with engine oil to the base of the thread. Any excess lubricant can be removed by inserting the finger. Refit the cork squarely in the orifice and wipe away any waste oil with a clean rag. *Author*

the return crank rod that connects with the return crank. The nut is retained with a taper pin, as with the nut on the trailing connecting rod, and must be examined and lubricated in the same way. On some later locomotives the return crank is fitted with a double taper roller bearing lubricated by a grease nipple. This type can be identified by a brass cover retained by nuts and bolts, beneath which the radius rod is still retained with the nut and taper pin; the cover is just to keep out the grit that would damage the bearings.

At the opposite end of the radius rod is the expansion link, pivoting on trunnions mounted in a bracket that forms part of the motion. The drive for the mechanical lubricator may be taken from the rear of the link, and is secured with a nut and split pin. Inside the expansion link is the die block, the supply of oil for which is carried in a cup containing a worsted pad that feeds drips of oil to the bearing surface. On the expansion link the trunnion pin bearings are filled in the normal way.

There are slight variations in the design of the expansion link of the Walschaerts valve gear, but the basic method for lubrication is the same, and with careful observation will present no problems.

When the last oiling point on the leading crank pin of the coupling rod has been lubricated, attention must pass to the little end bearing of the connecting rod. Here the oil may be carried in a cup in the end of the rod or in one fitted to the crosshead, from which the oil is fed by a worsted trimming to the oiling point. When the oil is carried in the end of the connecting rod and the locomotive is set in the correct position, the oil cup can be filled to its maximum capacity. Oil for the slide bars that control the passage of the crosshead is usually contained in an oil box, from where it is syphoned out to the lubrication point by a worsted trimming. It is advisable to place a smear of oil on the slide bars and piston rod themselves while the locomotive is standing. Some locomotives have a worsted pad fitted inside the bottom half of the crosshead that feeds oil directly on to the slide bar. There may also be oil reservoirs built into the crosshead that are fitted with corks and feed the oil by worsted trimmings.

Again there are differences in the design of slide bar and crosshead. Earlier types have slide bars top and bottom, while later designs used an arrangement with three slide bars at the top with the crosshead running in a groove. Just as with the expansion link, observation will resolve any problems with lubricating this point.

The slide bars are connected to the motion brackets with bolts and nuts secured with cotters; these must have their split ends open to keep them in place.

If the piston gland has been blowing there will be a trace of white emulsified oil on the front of the crosshead; this must be checked before the locomotive leaves the shed. Some packings are secured by two nuts, which can be tightened against the sleeved centre of the gland through which the piston rod passes to compress the packing round the rod. The nuts must be tightened one turn at a time each to keep an equal pressure on the soft inner packing.

While attending to the crosshead, attention must be given to the nut on the gudgeon pin, ensuring that it is tight and that the locking pin is in position.

Top and middle: The two methods of fixing the return crank of the Walschaerts valve gear to the crank pin. One is by means of four nuts and studs; to the right can be seen the connection with the rear connecting rod, with the nut, washer and cotter pin. The other method is the square with a cotter bolt; this must be checked for tightness when the locomotive is being prepared. *Author*

Bottom: The return crank/radius rod joint fitted with a double taper roller bearing; on top can be seen the grease nipple for lubricating the bearing. *Author*

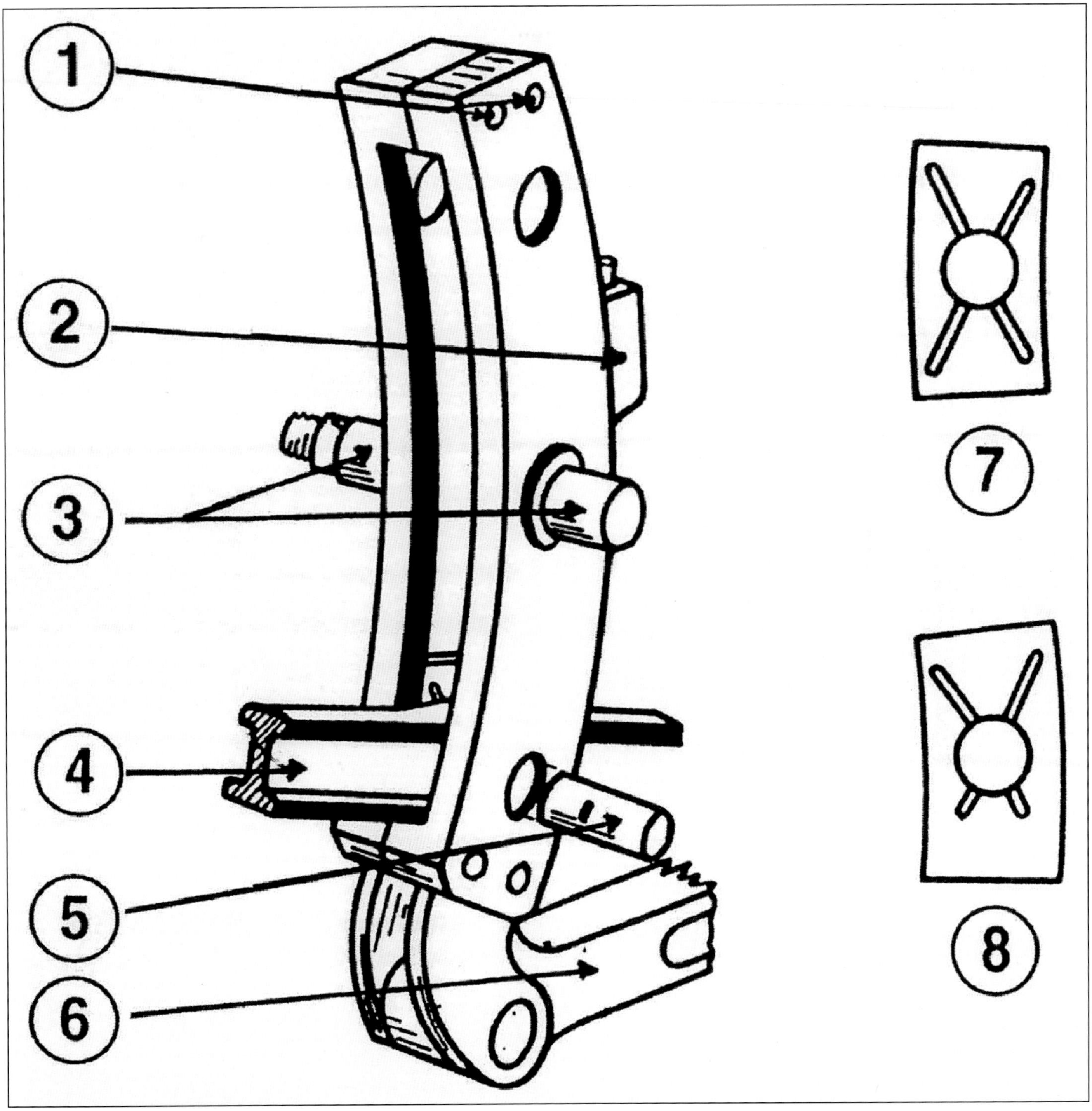

Diagram of a typical Walschaerts valve gear expansion link and die block: 1 countersunk bolts; 2 oil cup; 3 trunnions; 4 radius rod; 5 pin shown removed through hole; 6 eccentric rod; 7 oil grooves in side of right-hand die block next to expansion link; 8 oil grooves in side of right-hand die block next to radius rod

While working in this area, also place a few drops of oil on the pivot points of the cylinder drain cock linkages, using a small pressure oil can. The same process is now applied to the other side of the locomotive, finishing at the lubrication point opposite to that where the oiling started.

Later BR Standard locomotives made extensive use of grease nipples in several parts of the motion not previously noted for ease of maintenance, and this cut down the time required for preparation of the locomotive.

Three-cylinder Walschaerts

It is not possible to set a locomotive with three cylinders in one position to be able to lubricate the motion without re-setting the engine. The best method is to set the locomotive in the same way as with the outside motion; the only difference is that the crankpins are set at 120 degrees instead of 90 degrees, but it is possible to position the locomotive with the crankpins near to the top quarters with the coupling rod big ends nearly in the top back quarter to oil the outside motion

A close-up of the crosshead of a 'Black Five', showing the locking nut on the gudgeon pin secured by a castle nut and split pin. Just below the gudgeon pin is the anchor bracket from the crosshead to the union link. As with the adjacent combination lever, this incorporates oil points with sealing corks. At each end of the union link are the pins and collars that keep the pivot pins in position. On the top slide bar is the oil box that contains the lubricant for the crosshead and slidebars. Immediately under the oil box can be seen the cork for the little end bearing. Right at the top of the photograph is the valve spindle crosshead behind the oil feed pipe from the oil box on the main frame. *Author*

in the same way as the two-cylinder version.

After the outside motion has been oiled the locomotive can be moved so that it is over the inspection pit with the inside big end in the top back quarter; this will bring all the oiling points into a convenient position. Make sure before going under the locomotive that the wheels are chocked, the handbrake is on and 'NOT TO BE MOVED' boards are fitted prominently.

Stephenson link motion (outside connecting rods)

The valves of the Stephenson link motion are controlled by eccentrics running in sheaves. The motion of a locomotive fitted with outside connecting rods is best positioned in a similar way to the Walschaerts valve gear; if the crankpin on the left-hand side is in the top back quarter the eccentrics should be in a position where they can all be reached. The eccentrics fitted to the axle that control the valves between the main frames should also now be in a convenient position to reach with the feeder can.

With this type of locomotive it is usual to climb up into the narrow gap between the firebox and the axle to reach the oiling points. Clearly the importance of chocking the locomotive cannot be over-emphasised; other personnel should also be informed verbally that someone is working between the frames, and 'NOT TO BE MOVED' boards put in place.

Just as with the Walschaerts motion, the lubrication is carried out in a systematic manner starting at the trailing crankpin on the left-hand side of the locomotive and working round the outside. The oil in the oil cups is retained with corks and is treated in the same manner as the Walschaerts valve gear.

When the outside has been lubricated and the locomotive is secured by chocks, a start can be made on the inside motion; careful inspection will show the location of the oil points. The die blocks in the expansion links are lubricated with a worsted pad immersed in oil drawn from an oil cup.

Under the running plate of certain Great Western tank locomotives can be found an oiling point, normally on the main frame of the locomotive, for the tail rod of the piston valves; this can be filled with a special small feeder can.

Stephenson link motion (inside connecting rods)

Due to the design of this version the motion is cramped between the frames, and includes the connecting rods with their two big ends as well as the eccentrics for the valve gear. Only the coupling rods are outboard of the main frames.

The locomotive should be placed over the inspection pit to carry out the lubrication. Work round the outside of the locomotive, starting at the usual place with the rear coupling rods. If the locomotive is set with the motion on the crank axle in the recommended position, the oil cups should be easily reached. It is easier to go under the engine to carry out the oiling of the inside motion. As always, the engine should be chocked and 'NOT TO BE MOVED' boards fitted.

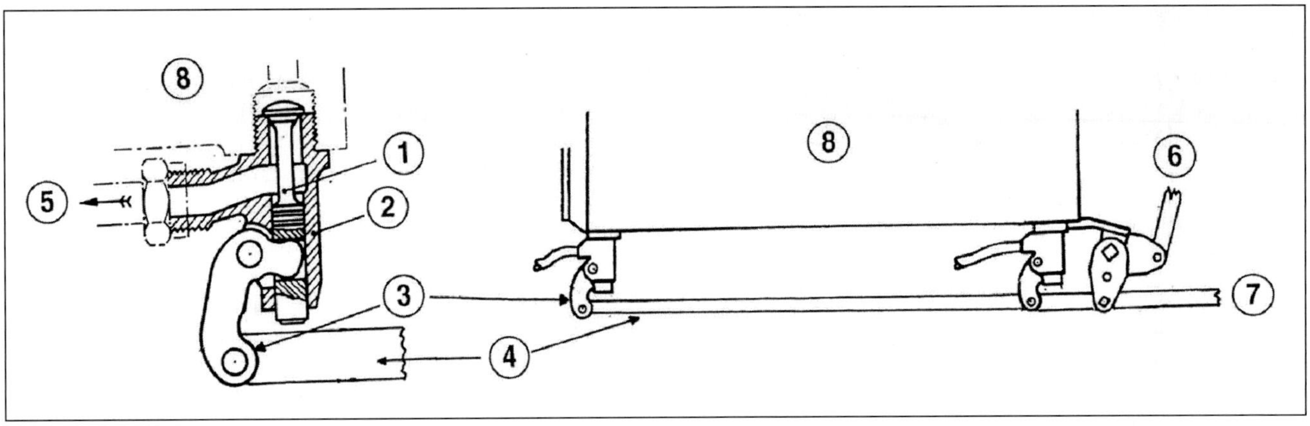

Diagram of the arrangement of the cylinder drain cocks fitted to the cylinder casting. Note the connecting link to the atomiser cock and the footplate control: 1 valve; 2 body; 3 lifting lever; 4 operating arm and connecting linkage; 5 to drain; 6 linkage to atomiser cock; 7 control from footplate; 8 cylinder

Bulleid chain-drive

The motion fitted to this type of locomotive is completely encased, running in a steel oil bath between the main frames. To maintain the oil level in the oil bath there is a large filling hole with a screw plug at the rear of the casing. When the oil level is correct the oil will be seen in the base of the neck of the filler. From there the oil is fed to the various parts of the motion by pumps, first through a filter then by pipes to oil cups where it drips on to the part requiring lubrication. Excess oil drops back to the oil bath and is re used.

Three additional mechanical pumps are fitted under the front between the frames just below the smokebox door, to lubricate the cylinders and the steamchests.

Oil to the axlebox guides is fed from worsted trimmings in large brass oil boxes on the footplate of the locomotive. Inside each box are separate sections that direct oil to the various lubrication points.

CHECKING AND FILLING OIL BOXES AND MECHANICAL LUBRICATORS

Oil boxes

Along the top of the running plates are the oil boxes (usually made from brass with a hinged lid) that connect with the horn guides and other sliding surfaces such as the pony truck, etc. As with the oil boxes on the motion, it is advisable to check them for any accumulation of water; as already mentioned, a cursory glance will not show the water because the oil will float on top, and it will be water instead of oil that is syphoned out of the oil box by the worsted trimming. The check can be carried out using a small hand syphon pump or syringe. Place the end of the pump in the bottom of the oil box and syphon out a little of the oil; if there is no or little resistance, it indicates the presence of water. Further evidence of water is provided by white emulsified oil being discharged from the pump when it is emptied.

Mechanical lubricators

Also on the main frames will be found the mechanical lubricators for the driving wheel axleboxes and cylinders. Each is clearly marked on the lid, since they employ different types of oil. Cylinder oil is thick to cope with the high temperature of the steam, and the lubricator incorporates a warming system, which must be turned on when the locomotive is being prepared so as to warm the oil before departure. The axlebox lubricator utilises normal engine oil, as no problem exists with high temperatures.

Some lubricators have hinged lids secured by a spring-loaded clip, while others have a screw filler. A firm pull will open the hinged lid to reveal a fine mesh filter or sieve below (see the diagrams in Chapter 1). The filter is in two halves, and when full the lubricant will be up to the filler level bar situated between them.

The oil must be poured on to the sieve and allowed to flow into the lubricator body, made easier when the thick cylinder oil is warmed beforehand to lessen its viscosity. The oil must never be poured directly into the main body of the lubricator without the sieve in place, or grit may find its way into the lubricator, causing damage to the feed pumps.

Both axlebox and cylinder lubricators work on exactly the same principle. The drive for the lubricator is taken from a convenient part of the motion such as the expansion link, so that it operates whatever the direction of the locomotive. The drive is delivered to the camshaft through a ratchet mechanism on the side of the lubricator body; the camshaft in turn drives the row of pumps that draw the oil through a series of ball valves. The pumps work independently of each other and are designed to meter 2oz of oil per 100 miles.

In order to make the cylinder oil flow the lubricator is fitted with a steam warming pipe that runs through the oil reservoir, and is controlled from a stop cock on the frames close to the lubricator body.

The cylinder oil then passes through the atomisers where it is turned into a creamy, emulsified liquid before being fed to the check valves at each lubrication point (see also the diagrams in Chapter 1). The steam supply to the atomisers is taken from the boiler and is controlled by a shut off valve; it is also connected to the cylinder drain cocks. When the cylinder cocks are closed the steam supply to the atomisers is turned on, and vice versa; it can therefore be seen that it is important

An oil box on the side of a locomotive's main frames with the lid open showing one of the worsted trimmings removed from the hole that connects with the feed pipe. This particular example feeds the slide bars to the valve spindle crosshead. Behind on the right is a mechanical lubricator, this one supplying oil to the cylinders, as clearly stated on its lid (see also the cut-away diagrams in Chapter 1). Behind the oil box are the atomisers, to which steam from the boiler is fed to emulsify the oil before it is passed to the cylinders. Behind the driver's wrist is the stop cock of the lubricator warming system, which warms the oil before it is fed to the pumps of the lubricator. *Brian Dobbs*

The top of a Silvertown cylinder lubricator with screw filler. On the side can be seen the oil feed pipes that run to the atomisers then to the cylinders. On the right-hand end is the ratchet mechanism with the handle for priming the system during the preparation of the locomotive. *Author*

Above left: A view of the inside of an axlebox lubricator. The two halves of the sieve can be seen, with the oil level bar between them. The pipes feed the oil to the driving wheel axleboxes on one side of the locomotive. *Author*

Top left: Refill the lubricator to the level on the sieve - never fill it with the sieve removed, or damage can be caused to the pump assemblies in the lubricator. *Author*

Middle left: After filling, the lubricator can be primed by turning the handle on the ratchet. Never exceed six revolutions of the handle or the flexible connections can be damaged, caused by hydraulic pressure. *Author*

Left: Fill the cylinder lubricator in the same manner with preferably warm cylinder oil to assist the flow - note the difference in thickness from ordinary engine oil. The lubricator can also be primed in the same way. *Author*

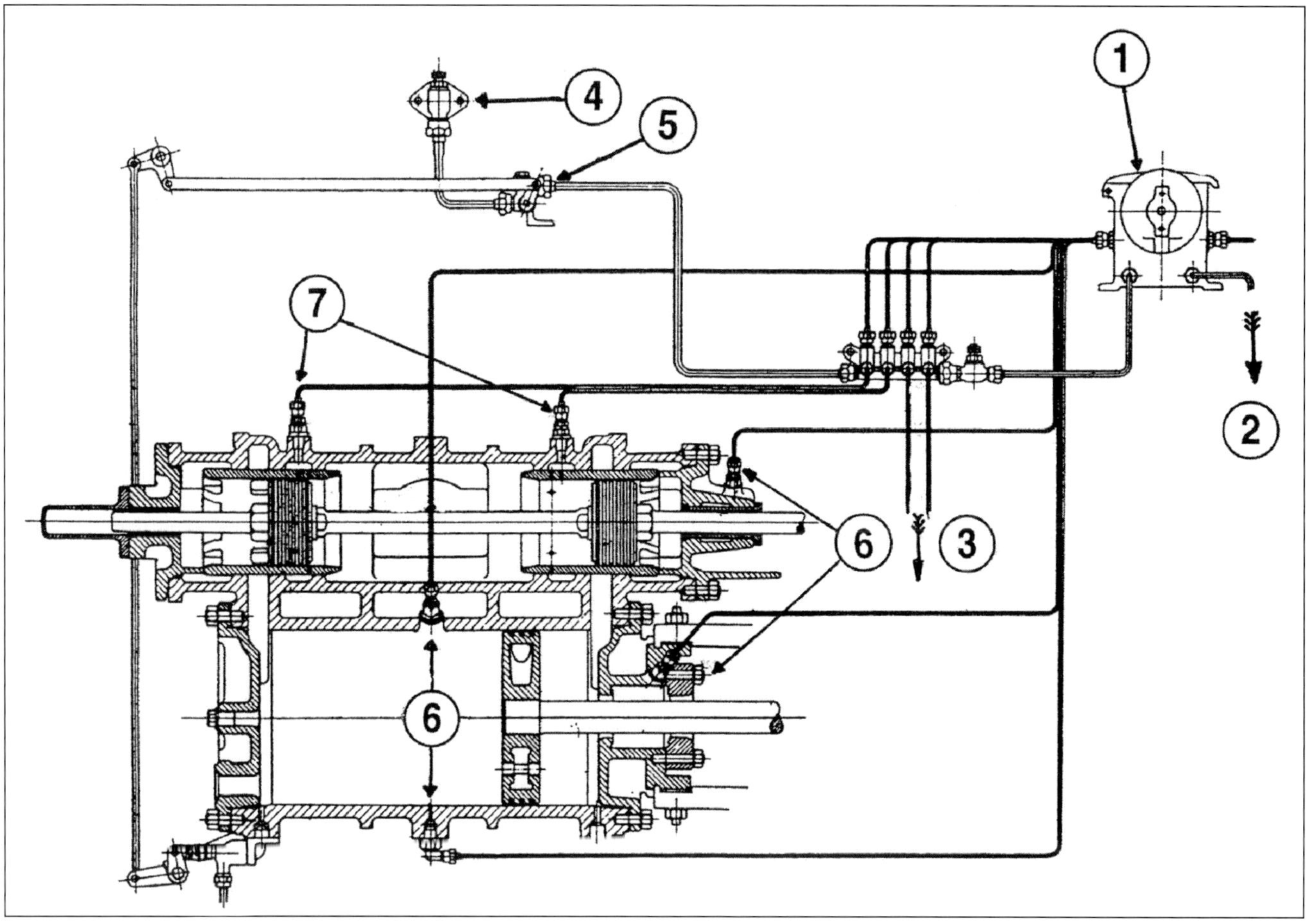

Above: Diagram of the layout of the cylinder lubrication system for a Midland-type locomotive. Note the shut-off valve for the atomisers on the smokebox side (4) and the automatic valve linked to the cylinder drain cocks (5), which is closed when the drain cocks are open: 1 mechanical lubricator, 2 lubricator warming pipe drain; 3 oil supply to other steamchest; 4 atomiser stop valve with tell-tale steam jet if valve is closed; 5 steam control cock for atomiser; 6 back pressure valve connections; 7 non-back pressure valve connections

Below: A back-pressure valve through which lubricant is fed to the cylinders and axleboxes: 1 body; 2 sieve; 3 ball valve; 4 spring; 5 bracket; 6 connection to flexible pipe

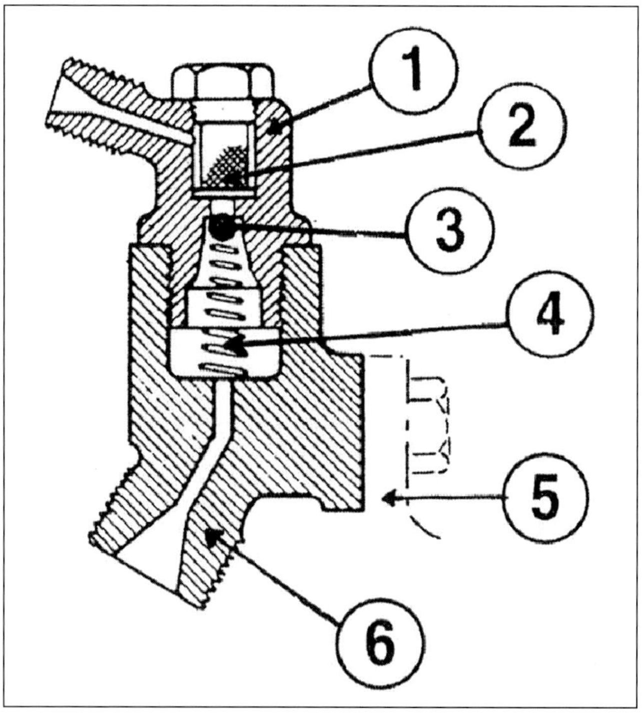

to open the cylinder drain cocks when the locomotive is standing for some time. In addition to the automatic shut-off to the atomisers, there is an additional stop cock that closes the steam supply from the boiler, and this must be opened when the locomotive is being prepared.

When the lubricator is full the system can be primed by turning the handle on the ratchet drive mechanism a few times; however, this must not exceed six revolutions as damage can be caused by the hydraulic effect to the flexible connections to the axleboxes.

Hydrostatic lubricators

This type of displacement lubricator is fitted to most Great Western locomotives for the lubrication of the cylinders (see also the diagrams in Chapter 1), and it is usual for the fireman to re-fill it as part of his duties when he is preparing the locomotive. With the larger Great Western locomotives such as the four-cylinder 'Castle' and 'King' classes, with their many oiling points, the preparation time for oiling the locomotive is tight, as there are well over 100 separate places, some in the most inaccessible locations.

If the lubricator has been re-filled when the driver returns to the footplate after oiling round, he should turn on the warming cock to heat the oil before he starts filling the top boxes. This will give the oil time to warm through before it sets and adjusts the flow prior to the engine leaving the depot. The lubricator can remain closed while the warming takes place; the steam from the warming cock passes through a radiator in the rear of the oil reservoir in the lubricator body, escaping into the drain pipe where the water is drained from the body.

Once the oil is warm the warming cock can be turned off; the body of the lubricator should be hot to the touch. The lubricator can now be started and the flow adjusted. This is carried out by first opening the central stop cock between the manifold and the condensing coil in the cab roof. The coil is in two separate sections, one for each side of the lubricator, and the stop cock can be identified by the square nut with the pointer that indicates which side of the coil is being used. When the pointer is vertical the steam supply is closed; to each side it points towards the coil being used; and when pointing down steam is being fed to both coils.

On top of the lubricator body is a shut-off valve that connects with either of the condensing coils. When this has been opened each valve on the centre oil gallery at the front of the lubricator is opened and the oil flow adjusted to one drop per 30 seconds by means of the sight glass on the front of the lubricator body. The start-up procedure and flow adjustment is the same for all Great Western locomotives.

At the side of the lubricator there is a knurled adjuster, which passes oil directly to the regulator valve only; this must be open all the time and the oil feed adjusted to suit.

AXLEBOXAND HORN GUIDE LUBRICATION

Driving wheel axleboxes (LMS type)

If it has not already been done in order to reach the inside motion, the locomotive will need to be placed over the inspection pit to reach the axleboxes and horn guides.

Inside the axlebox is the oil bath, also known as the underkeep, which forms the bottom part of the axlebox. In the oil bath is a semi-circular steel frame carrying a worsted mop pad that is kept in contact with the bottom axle journal by a compression spring and syphons oil from the bottom of the oil bath. As the axle rotates the pad leaves a film of lubricant on the journal.

Every day before the locomotive enters service the axleboxes must be drained of any water that has built up in the oil bath. The water finds its way in either from steam condensing around the locomotive or, in the case of a tank engine, when the tank is allowed to overflow when filling.

The underkeep of the axlebox in the horn guide, seen from between the frames (see also the diagram in Chapter 1). Fitted to the horn guide is a frame stretcher that connects with the horn guide on the opposite side of the locomotive. The underkeep contains the worsted mop pad; at the bottom is the drain bolt, and just above it the filler cork, while to the left is the main oil feed to the axlebox. On the face of the axlebox is the casting that contains the oil seal; this requires no attention by the driver except to check that no large oil leak exists, indicating a faulty seal. *Author*

Left: Another view showing the spring arrangement connected to the axlebox and the two nuts, secured with split pins, on one end of the frame keep. *Author*

Above: A pony truck axlebox showing the filler. Not visible is the hexagonal drain bolt at the front bottom. *Author*

If the water was left to accumulate in the underkeep it would eventually replace the oil. The mop then would just syphon water to the journal, causing the axle to run hot. This very serious defect is known as a 'hot box' and can be first heard as a high pitched squeal from the axle when the locomotive is in motion. Further evidence of the fault can be found by gently feeling the end of the axle in the wheel boss with the back of the fingers; it will be very hot to the touch. Sometimes during the day, possibly while the fireman is taking on water, the driver will walk round his locomotive systematically feeling the axle ends for early signs of a hot box.

The task of draining the axleboxes is quite simple to carry out. The drain bolt fitted to the axleboxes of ex-LMS locomotives can be found in the neck just below the filler cork, which should be removed so that air will enter the axlebox when the bolt is slackened. The bolt is designed so that when it is released by a few turns any water that has built up in the axlebox will be drained out of the oil bath until oil can be seen running from the box. If nothing at all appears, it could indicate that the drain hole is obstructed, which will necessitate the removal of the drain bolt.

There will usually be no more than a few teaspoons-full of water found in the oil bath, and when it has been drained the cork can be removed and the oil level checked, topping up if necessary with engine oil.

Driving wheel axleboxes (Great Western type)

The axleboxes fitted to the driving wheels of Great Western locomotives carry their lubricant in oil baths contained on the top of the axlebox and in the underkeep below the axle. The level of lubricant in the axlebox is maintained by manual filling, pouring the oil through the filler fitted with a cork to the top and bottom of the axlebox.

The task is made easier by once again warming the oil before filling. As there is no level mark on the axlebox this is a good guide as to the level of lubricant.

Certain other types of Great Western locomotives have their lubricant fed from a box on the side of the frame.

Horn guides

While under the locomotive it is important to place a few drops of oil in between the axlebox and the horn guide, to prevent the axlebox from sticking. This can help to prevent the locomotive from slipping on a bad rail; if the axlebox partially seizes in the horn guide it will put the locomotive out of balance by holding the driving wheel fractionally above the rail; it could also easily lead to a derailment. If the surface of the horn guide appears dry, use hot oil, which will be thinner and thus flow more easily on the surface. The same method of lubrication applies to the horn guides on the pony truck and Bissel truck.

Pony truck axleboxes

The axleboxes of the pony truck are not usually fed by the mechanical lubricator, so the oil level must be replenished by hand with the feeder can. These axleboxes can be treated in a similar way to those of the driving wheels, except that the drain bolt will usually be found in the bottom of the under keep.

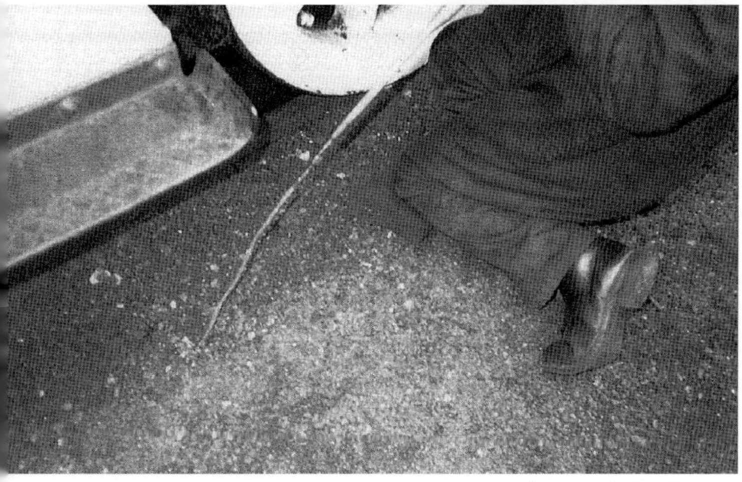

Top: The metal lid on top of a well-type axlebox, held open by a wooden wedge while maintenance is carried out. *Author*

Above: Draining the well-type axlebox: a syphon pump is inserted below the metal lid and any water syphoned from below the oil, then exhausted from the pump. Water will be easily pulled into the syphon pump giving little resistance. When the axlebox well is empty it can be refilled with engine oil, again using the syphon pump. *Author*

Well-type axleboxes

This type of axlebox is usually fitted to industrial locomotives that travel short distances, mainly round goods yards and stations, although it may be fitted to older main-line locomotives. The oil is carried in a reservoir on top of the axlebox, usually divided into two sections, from where it is syphoned by the worsted trimming and gravity to a felt mop running in contact with the bearing.

When preparing a locomotive with this type of axlebox it will, as usual, be necessary to first check the axlebox for water, which must be drained out before re-filling is attempted; this is done in the same way as described above, taking care not to disturb the trimming from the feed hole that delivers the oil from the reservoir to the bearing. These boxes usually have a metal lid that sits on top of the axlebox, which has to be lifted with a large screwdriver to gain access to the reservoir; the screwdriver can then be left to support the lid while maintenance is carried out. After the box has been checked for water and drained as necessary, it is re-filled with normal engine oil until the oil is seen to run out.

Tender axleboxes (LMS type)

There are two types of axleboxes fitted to former LMS-type tenders, although they are basically the same design.

The first type has an axlebox cover secured on each side by a nut and bolt; the nuts can be slackened but will not come off by accident due to being retained on the thread with split pins. One of the lugs on the axlebox cover is slotted, so when the nuts are slackened the cover can be rotated on the other bolt. In the centre of the cover is a hexagonal nut that will accept a 7/8 inch spanner; this is used to turn the lid to the open position.

At the bottom of the axlebox will be seen the lubricant; again this will have to be checked for water at the bottom and re-filled with oil if necessary. A visual check of the worsted mop can also be made while the axlebox cover is opened.

When the axlebox is closed care must be taken not to trap the end of the frame that carries the mop in the cover.

On the later type of axlebox the oil can be simply checked and re-filled through a cork at the side. Draining of any water from the axlebox is carried out as on the earlier type.

The horn guides can also be lubricated while the axleboxes are being checked for oil. This is best carried out with a pressure oil can, directing a small amount of oil between the axlebox and guide.

Roller bearing axleboxes

Taper roller bearing axleboxes can be found either inside or outside the main frames of the locomotive or tender. In the case of inside mounting it is known as a cannon box; the roller bearings are mounted in a tubular cast steel casing that completely surrounds the axle. As with the older type of axlebox, vertical movement is controlled by springs, with the cannon box in between the horn guides. Inside the axlebox is a cone shaped bearing surface on which the taper roller bearing runs. Taper roller bearings are usually found on the tender and the Bissel trucks of British Railways Standard locomotives, but can also be found on the driving wheels of some later BR locomotives.

An LMS-type axlebox, showing the two bolts and nuts that secure the cover. These have been slackened and the spanner is on the centre nut ready to rotate the cover to open it. *Author*

Inside the axlebox can be seen the circular end of the axle journal, and on top of it the axlebox bearing. At the bottom are the two tabs at the ends of the worsted mop that syphons the lubricant to the end of the axle. The spanner is on the drain nut, loosened to drain any accumulation of water from the oil bath. *R. G. Fox*

Top: A tender axlebox on a 'Black Five'. Note the cork in the bottom corner for checking the oil level in the box, and the horn guide that controls the vertical movement of the axlebox under the spring. *Author*

These bearings require no attention on the part of the driver during the preparation of a locomotive; any maintenance is carried out by the fitting staff.

STEAM BRAKE, VACUUM PUMP AND COMPRESSOR LUBRICATION

LMS and BR Standard steam brakes

The lubrication point for the engine and tender steam brake will be found on the footplate. Because the brake is in contact with steam it is necessary to use cylinder oil; as the early type is filled through a hole closed by a bolt, filling will again be made easier if the oil is warmed before use.

The supply of steam to the brake will have to be shut off at the stop cock; on former LMS locomotives this is usually *a* ½-inch square nut on the manifold. There are three types of oil cups or reservoirs fitted to these locomotives. The older type has a bolt at the top on the side of the reservoir, which is slackened to open the filler hole in the top of the cup. Slowly pour the hot oil from the feeder can into the filler or use a pressure oil can until the cup is full. The later type has a screwed cap on the top of the reservoir with a filter inside the cup. Some of these oil cups have a drain bolt on the side at the bottom to remove any water from the reservoir. When the bolt is slackened the water will be seen to run from the cup; when it has drained, the bolt is re-tightened.

BR Standard locomotives have a similar filler with a screw cap, and it is usual to find separate fillers for the engine and tender steam brakes. The 9F class heavy freight locomotives have three steam brake cylinders and consequently have three oil filling points on the faceplate of the locomotive near the driver's brake valve.

The Great Western steam brake

The steam brake lubrication filler is to be found on the steam pipe leading from the brake valve to the brake cylinder, and can be identified by the screw plug with the bar on top, and usually a chain to prevent its loss. With the filler removed, cylinder oil can be poured down the pipe to lubricate the piston in the cylinder. Replace the screw filler and tighten.

Southern Railway steam brake

The filler for this type of steam brake is similar to that of the Great Western type, and is filled via the filler plug fitted to the steam pipe to the brake cylinder.

When the steam brake has been lubricated it will have to be turned on at the shut-off cock. It is advisable to make a visual test of the operation of the brake before the locomotive is used. This can be carried out if the locomotive is chocked so that the hand brake can be taken off; this must be done with a tank engine where the hand brake operates on the driving wheels. If the fireman operates the steam brake lever the

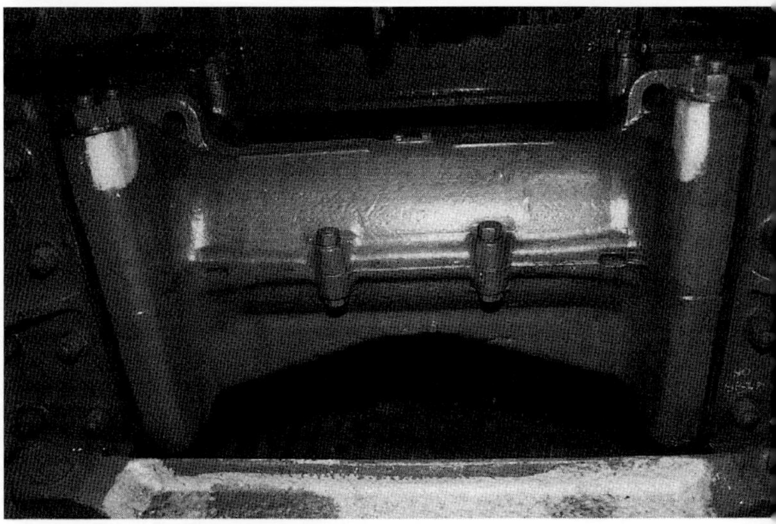

Middle: Another 'Black Five' tender axlebox with an oilbox fitted to the top plate. This plate is not attached to the axlebox but allows it to slide vertically in the horn guide. *Author*

Bottom: A cannon-type roller bearing axlebox situated between the frames, with the horn guides on the outer edges of the casting. At the front can be seen the stretcher running between the two axleboxes. *R. G. Fox*

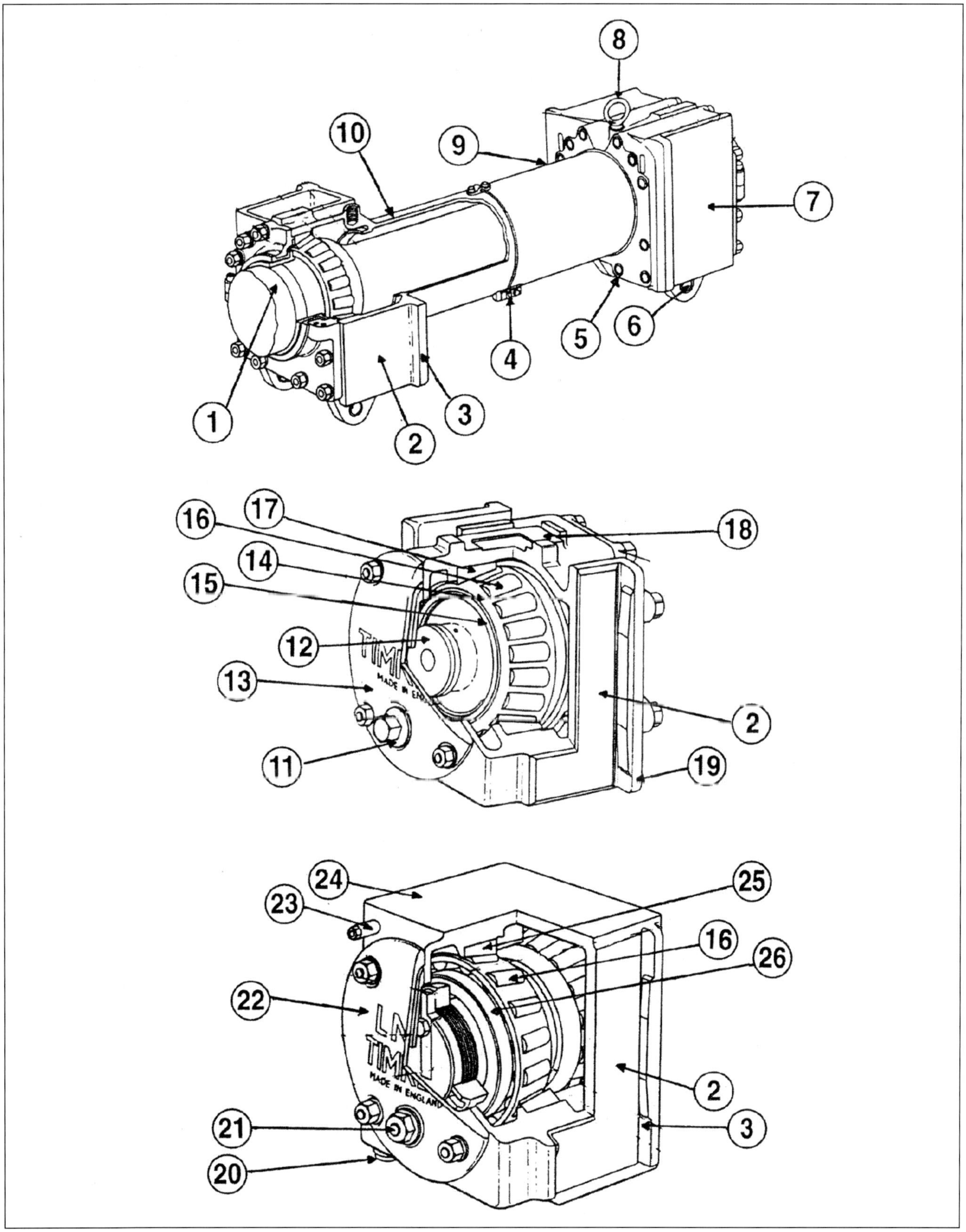

Diagrams of typical roller bearing axleboxes. At the top is the double-row type containing two rows of taper bearings. The centre example has a single row of taper roller bearings, while at the bottom is a cross-section of a cannon box as fitted to a driving wheel axle inside the frames: 1 wheel seat; 2 horn guide; 3 horn guide lug; 4 air vent; 5 oil drain plug; 6 spring hanger pin hole; 7 axlebox; 8 lifting eye; 9 oil filler plug; 10 tubular steel housing; 11 oil filler plug; 12 axle; 13 inspection plate; 14 cone; 15 cage; 16 rollers; 17 taper cage; 18 spring seat; 19 back cover; 20 oil drain; 21 oil filler plug; 22 inspection plate; 23 enclosed grease feed; 24 spring seat; 25 cup; 26 cone rib correct operation of the brake mechanism can be observed.

While the steam brake is being tested a check can also be made on the travel of the piston rod. Some steam brake cylinder pistons are what is known as a semi-plug type and operate in the following manner. The crown of the piston is in two segments separated by a taper piston ring. When steam is admitted to the cylinder the two segments are brought together, causing the taper piston rings to be forced outwards against the bore, making it steamtight and forcing the piston rod downwards to the end of the stroke. The action of the piston rod on the bell crank, or two-to-one lever, pulls the brake stretcher that applies the brake blocks to the wheels. The reason for this semi-plug type piston is to allow the brake to be released quickly with the assistance of the return spring. In order for it to work efficiently there must be sufficient free play in the rod.

Great Western brake vacuum pump

The vacuum on Great Western locomotives is maintained by a vacuum pump driven from the right hand crosshead, which will maintain the required level of vacuum at a line speed in excess of 15mph.

The pump is found under the running plate (see also Chapter 1), and lubrication is carried out in the following manner. On top of the pump body can be found a screw-type stopper, which is removed with a spanner. The pump is then re-filled with a mixture of 50 per cent oil and 50 per cent paraffin; this is used to thin the viscosity of the oil and allow it to flow more freely and not obstruct the pump's valves.

In the case of a tank locomotive the oil is carried in a separate container mounted at the side of the water tank.

The Westinghouse air compressor

The lubrication point for the air compressor can be found on top of the pump body (see also the diagram in Chapter 1). Because of the temperature it is necessary to use cylinder oil in the reservoir. At the bottom of the oil cup there is a drain nut to remove any condensation that may build up.

Ensure that the steam is shut off to the pump before draining any water from the cup. Loosen the drain nut and filler cap, then, when the water has run out, re-tighten the drain nut and re-fill the reservoir to the top of the neck. Replace the filler plug and tighten.

FILLING AND MAINTAINING STEAM SANDERS

Adhesion problems have always existed with railway locomotives due to the metal-to-metal contact between the rail and wheels; the stress caused by a slip of the wheels can cause damage to the motion of the locomotive, and excessive slipping can also cause priming, thereby inflicting more damage to the locomotive.

To eliminate this problem a steam locomotive is fitted with devices to apply sand to the rails in front of the wheels, and several methods were tried before steam-operated sanders came into popular use. Steam at boiler pressure is admitted to the sand gun, which is mounted close to the rail just in front of the driving wheel; this creates a partial vacuum in the sand gun, which draws the sand from the sand box through a sand trap, down the pipe to the sand gun and out between the wheel and the rail.

The Timken taper roller bearing axlebox fitted to the Bissel truck of BR Standard 'Pacific' No 71000 Duke of Gloucester. Note the use of the dual coil springs instead of the more conventional leaf springs, in conjunction with the large beam keeping the spring in contact with the axlebox. *D. Dyson*

It is usual for a locomotive to have front and rear sanders that operate separately. A locomotive with six or eight driving wheels will be expected to spend most of its time travelling in a forward direction, so there will be four front-facing and only two rear facing sanders.

The sand boxes, which can be found on the main frames, should be checked and re-filled as part of the locomotive preparation procedure. When they have been filled they should be tested by the fireman by opening the valve on the footplate. This is a dual valve- one side operates the front or leading sanders and the other the rear or trailing sanders.

After the first drips of water caused by condensation, steam will be emitted from the sand gun; if the sander is working properly and the fingers are placed near the wheel in the jet of steam, grains of sand will be felt. If they are not, it could be that the sand boxes are empty, or that the sand trap is blocked with wet sand.

The blockage can be cleared by the following method. The sand trap is located at the base of the sand box, and is kept

Above: The early type of engine steam brake oil reservoir with the side bolt filler. A similar type may be fitted to the tender. Because cylinder oil is used, it will have to be warmed first. *D Dyson*

Right and below: Filling the later type of steam brake oil cup. First shut off the steam supply at the connection with the manifold, then remove the top filling cap as shown. Withdraw the gauze filter and check for water in the cup; if any is found remove the drain nut from the body to allow it to drain out. With the drain nut and filter replaced, the cup can be re-filled and the cap replaced. *R. G. Fox*

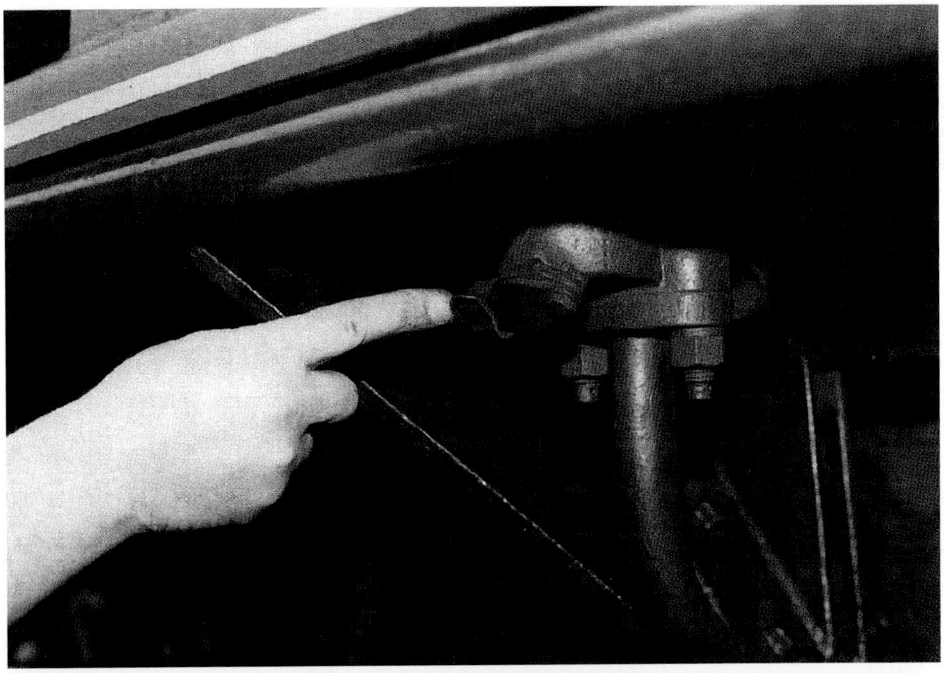

Steam sanders must be checked so that the sand trap at the bottom of the box is not filled with wet sand. When the small door in the trap is opened, sand should be seen to flow freely. If not it can be cleared by forcing a stout piece of wire into the trap until the sand flows. Sometimes the door may be replaced with a cork. *R. G. Fox*

closed by a hinged flap or cork. Open the flap or remove the cork and force a piece of strong wire, about 12 inches long, up the trap into the sand box until sand can be seen to run freely from the trap, then the flap can be closed and the sander tested once more.

FINAL CHECKS

Checking for steam leaks

When the oiling has been completed and the locomotive is in steam, a check can be made on various parts of the locomotive such as for steam leaks in the main internal steam pipes in the smokebox. With the locomotive in mid-gear, the wheels chocked and the handbrake firmly applied, open the smokebox door; if the fireman then opens the regulator handle, any steam leaks will be seen blowing from the joints. Such leaks will cause the smokebox vacuum to be partly destroyed, making the locomotive hard to steam. More seriously, if the packing of a steam joint or the steam pipe itself was to fracture in the smokebox, it would cause a disastrous blow-back of fire on to the footplate.

The next things to check are the cylinder piston and valve spindle glands. With the locomotive still chocked, place it first in forward gear, close the drain cocks and open the regulator slightly; if there is a leak, steam will be seen passing from the gland. Next place the locomotive in reverse gear and carry out the test on the opposite side. If a gland is blowing, it can be rectified with some types by tightening the nuts on the gland. If the blow of steam is serious a fitter must be called to carry out repairs. A blowing gland will cause the oil to waste from the cylinder and escaping steam will lead to poor visibility when the locomotive is travelling forwards.

Testing the vacuum brake

With a sufficient level of steam, the operation of the vacuum brake can be tested. First open the small ejector and the required level of vacuum will immediately be created. Close the ejector and destroy the vacuum by applying the vacuum brake, then, with the small ejector still closed, re test with the large ejector, when the vacuum will be seen to rise again. The large ejector will over-create vacuum, and air will be heard to enter through the vacuum adjuster (sometimes called the pepperpot) on the brake valve or connected to the train pipe near the vacuum brake valve.

Air leaks in the train pipe can be detected by creating the vacuum, then using a duck lamp (one that burns rape oil giving off a crude flame); the flame will be drawn by the airstream towards the train pipe.

Testing the exhaust steam injectors

Before leaving the shed to enter traffic with a locomotive fitted with exhaust steam injectors, it will be necessary to test the automatic change-over valve that changes between live steam and exhaust steam operation. This can be carried out in the following manner. The locomotive should be chocked with the handbrake applied, and the injector started. At first it will be working on live steam. If the regulator is then opened the shuttle valve will be actuated and water will be seen emitting from the overflow; as the locomotive is stationary the injector will not be able to start because there is no exhaust steam passing from the cylinders. Now close the regulator and open

The traditional enginemen's white enamel tea can with a detachable lid that forms a drinking cup. This is the best method of keeping tea or coffee warm on the footplate. (Beesley Products)

Bacon being fried on the shovel inside the firebox of 'Black Five' No 45337 after preparation of the locomotive. The coal must be well burned through and not giving off smoke, or this will flavour the food. Note the can of tea on the warming plate over the fire door. *Author*

the cylinder drain cocks to relieve the pressure on the shuttle valve - the injector will then re-start on live steam.

Reporting defects

Any defect must be reported to the fitters before the locomotive departs from the depot. They may decide that the engine is safe to enter traffic, and the repair will be effected on its return.

When all the above checks and work have been carried out, the locomotive can be passed as ready for departure into service.

COOKING ON THE FOOTPLATE

After the preparation has been completed the reward of a good breakfast cannot be underestimated; it is another very good reason for an early start to leave time to enjoy the meal! Just as the preparation work on the locomotive is planned, so is the cooking of the breakfast.

The meal can be eaten from the shovel, but it is preferable to use a tin plate. It is essential that the plate is well warmed beforehand, either over the warming plate or in some other suitable place. Depending on the type of food planned or the number of people to be catered for, the tin plate will be extremely useful.

During the preparation of the locomotive the fireman will need to level the fire in order to raise steam; if a fried breakfast is planned, this should be done early to allow the fire to burn through with no smoke being given off, which would otherwise seriously affect the flavour.

The shovel should have no large pits in the metal surface, which could cause burning of the food. Wash the blade of the shovel with the slacking pipe or in a bucket of hot, clean water, then dry it with a clean rag or paper towel. Warm the blade over the firebed, then, when it is hot and dry, remove it from the firebox and tilt it backwards to form a 'pan' in the base of the shovel close to the handle.

Various types of food can be cooked on the shovel, including bacon, sausages, eggs, steaks, ham, fried bread, black pudding, etc. It is possible to cook the meat together with the eggs, or do the eggs separately. Place a generous amount of lard or cooking oil into the deep part of the shovel and allow it to melt. Place the meat in the melted fat, then move it up the shovel a little and break the eggs into the pool of hot fat. Replace the shovel over the hot firebed but well clear of the fire, or the hot fat might ignite. Turn the food frequently until it is fully cooked, then remove the shovel from the fire and place the food at the end of the blade to allow the excess fat to drain back into the base of the shovel.

If the eggs are being cooked separately, place the drained meat on the hot plate. Leave the shovel tilted backwards to retain a reservoir of hot fat, which should still be hot enough to cook the eggs. Break the eggs into the fat and allow them to cook, basting them with the fat and replacing the shovel over the firebed if the fat is not hot enough.

Pasties and pies can also be warmed through on the warming plate. Completely wrap them in aluminium cooking foil, then place them in a clean newspaper outer jacket. Finally place the contents in a polythene bag to protect from any oil or water that may fall from the regulator gland packing. The newspaper is used in case the heat from the fire melts the polythene bag. Although the cooking time will vary depending on the size of the pie and the type of locomotive, they usually take about an hour.

Tins of soup or rice pudding can be warmed through by piercing a hole in the rim of the can and placing it in a convenient recess on the faceplate.

Pies can be warmed on the warming plate, and will take about 1 to 2 hours depending on how the locomotive is being worked. First they are wrapped in cooking foil, then in a newspaper parcel, and finally in a polythene bag to protect the food from contamination by drips of water, etc., from the faceplate. *Author*

6
Locomotive controls

Although they vary in design and operation from company to company, the controls of British locomotives are basically the same, performing the same tasks. On the following pages are representative footplate layouts from the former 'Big Four' companies and a BR Standard example.

The footplate of an LMS Ivatt Class 2 locomotive. *R. G. Fox*

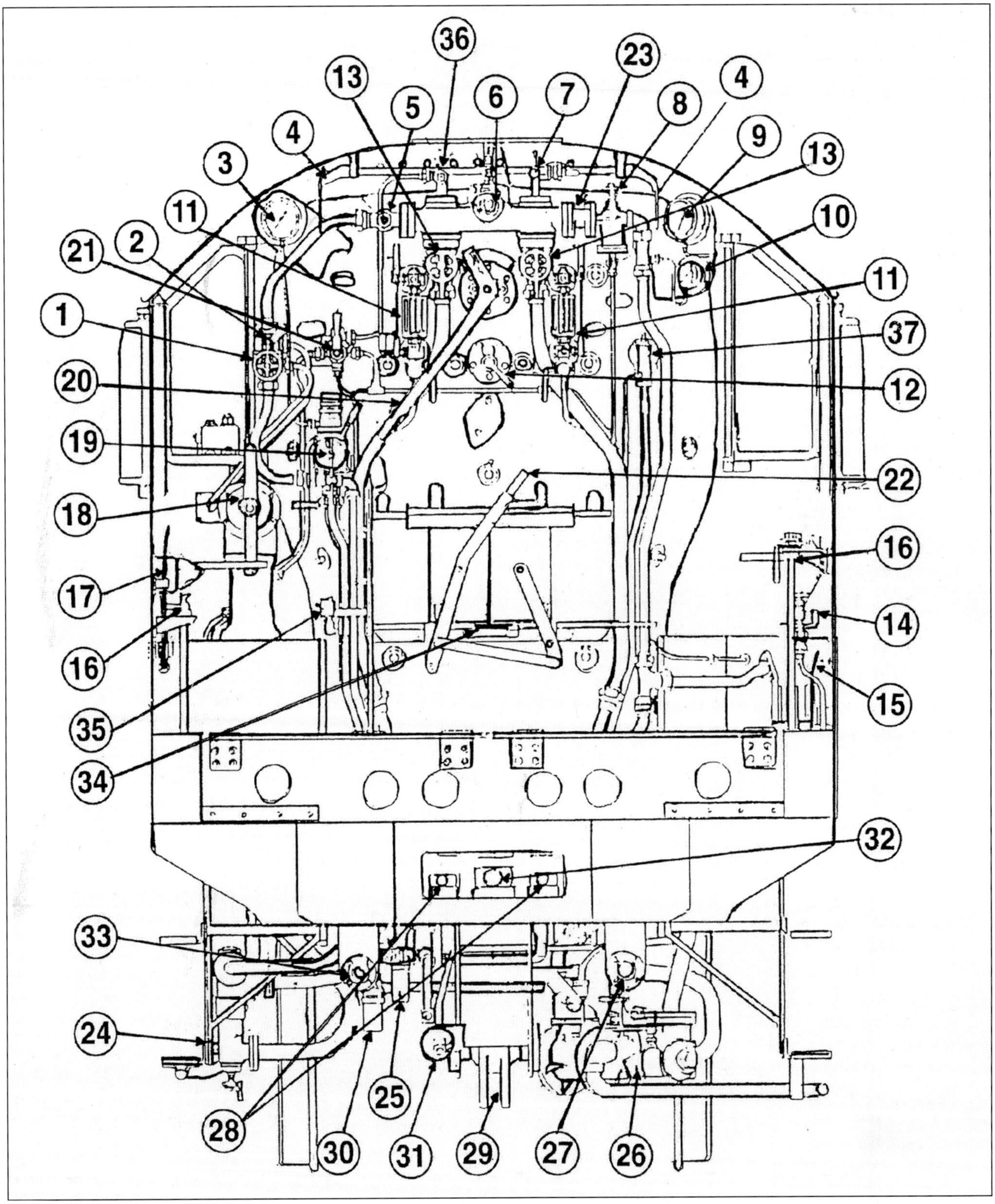

Layout of a typical LMS footplate: 1 large ejector steam spindle; 2 small ejector steam spindle; 3 vacuum gauge; 4 whistle; 5 steam brake shut-off cock; 6 manifold shut off; 7 whistle valve; 8 carriage warming regulator; 9 steam pressure gauge; 10 carriage warming pressure gauge; 11 boiler water gauge; 12 blower valve; 13 injector steam spindle; 14 slacking pipe control; 15 damper controls; 16 injector water regulator handle; 17 cylinder drain cocks; 18 reverser; 19 combined steam and vacuum brake; 20 regulator handle; 21 sanding control; 22 firebox door handle; 23 carriage warming; 24 live steam injector; 25 carriage warming pipe connection to tender; 26 exhaust steam injector; 27 water connections; 28 small drawbars; 29 steam brake cylinder; 30 carriage warming drip valve; 31 blowdown pipe to tender; 32 main drawbar; 33 steam brake connection to tender; 34 lip plate; 35 steam brake oiling cup; 36 sanding steam shut-off; 37 continuous blowdown valve

The footplate of Great Western Collett 2251 class 0-6-0 locomotive No 3205. Note the steam brake and vacuum brake arrangement with the diaphragm above the brake valve. The locomotive is fitted with a three-feed lubricator. The rest of the controls are similar to other Great Western locomotives. *Peter Herring*

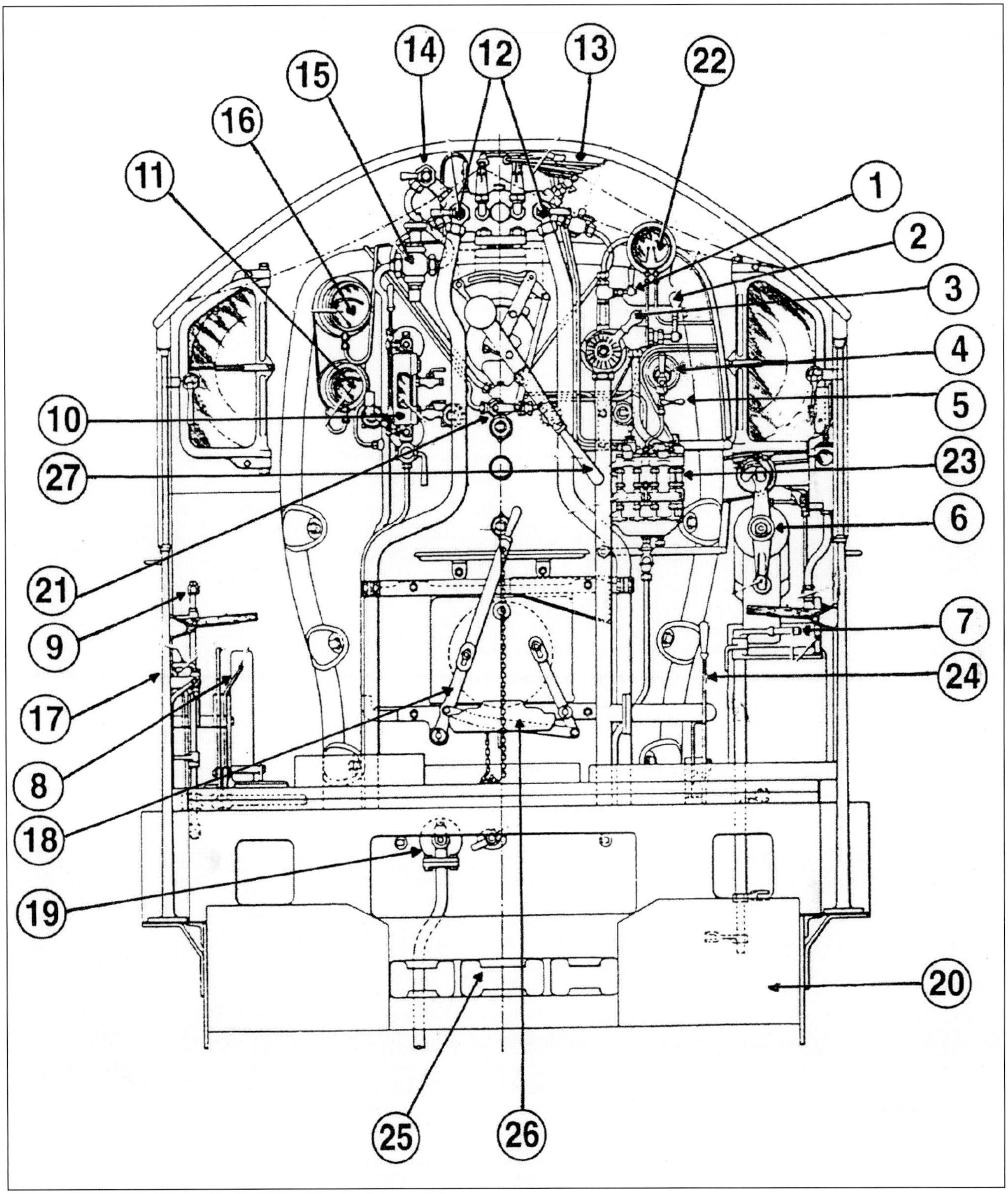

Layout of a Great Western 'King' class footplate: 1 small ejector; 2 large ejector; 3 vacuum brake; 4 blower valve; 5 lubricator warming cock; 6 reverser; 7 sanding gear; 8 damper controls; 9 exhaust injector control; 10 water gauge; 11 carriage warming pressure gauge; 12 injector steam spindles; 13 lubricator condensing coil; 14 carriage warming cock; 15 carriage warming regulator; 16 steam pressure gauge; 17 slacking pipe control; 18 firebox door handle; 19; blowdown valve; 20 intermediate budding gear; 21 lubricator jigger valve for regulator; 22 vacuum brake gauge; 23 lubricator; 24 cylinder drain cocks; 25 drag box; 26 firebox door flap plate; 27 regulator handle

The footplate of Southern Railway 'King Arthur' class 4-6-0 No 30777 Sir Lamiel. *Brian Dobbs*

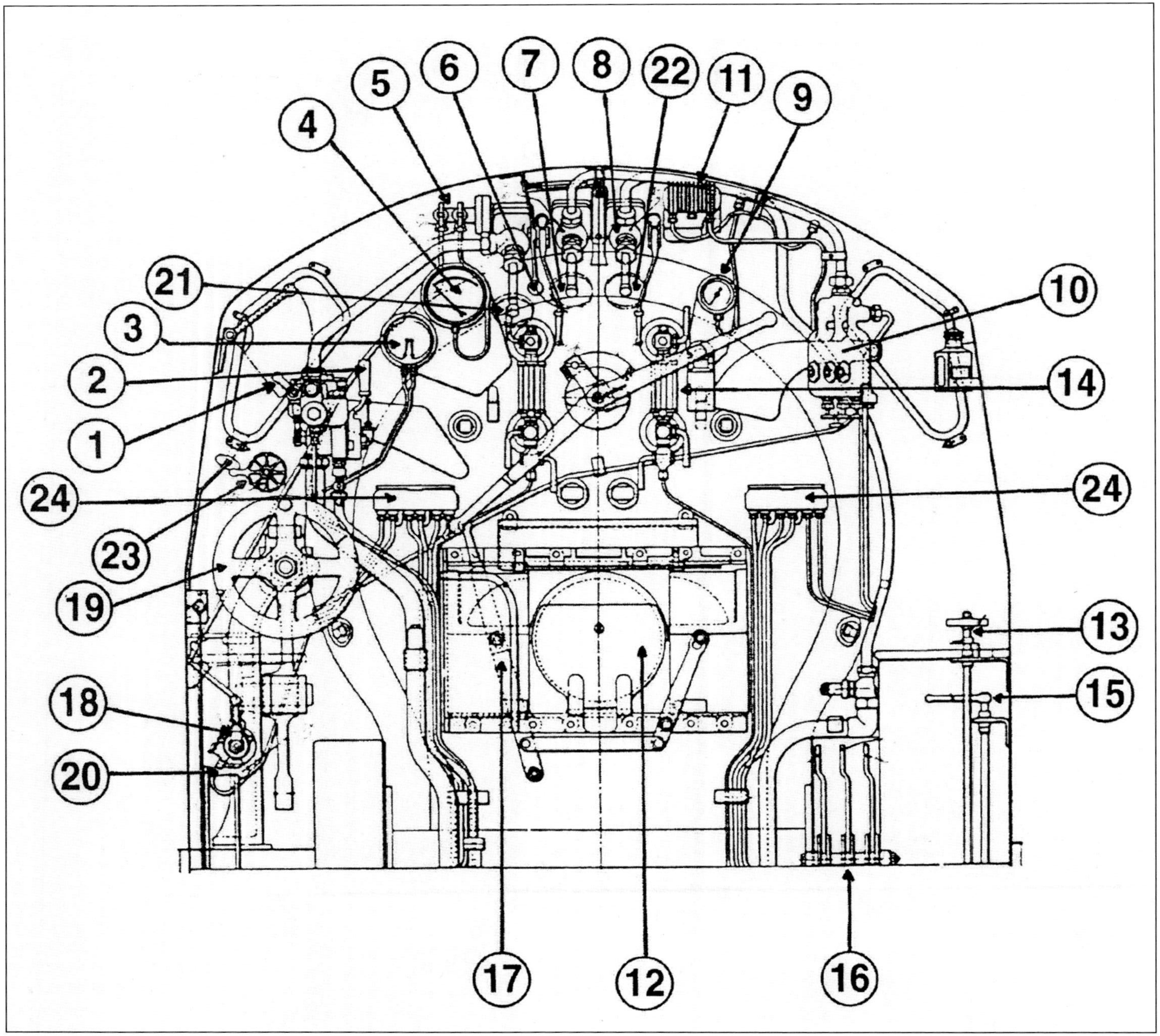

Layout of a Southern Railway 'Schools' class footplate: 1 small ejector; 2 driver's brake and large ejector lever; 3 duplex vacuum gauge; 4 pressure gauge; 5 steam shut-off cocks to clutch and pressure gauge; 6 whistle lever; 7 sight feed lubricator steam valve; 8 train heating steam valve; 9 steam heating pressure gauge; 10 lubricator; 11 condensing coil; 12 fire door flap; 13 water regulator (exhaust steam injector); 14 water gauges; 15 sanding lever; 16 ashpan levers; 17 fire door lever; 18 reversing shaft clutch valve; 19 reversing wheel; 20 cylinder drain cocks; 21 ejector steam shut-off; 22 vacuum release valve; 23 blower valve; 24 oil boxes

A London & North Eastern Railway locomotive footplate - this is a B1 class 4-6-0. As can be seen, the locomotive is fitted with a steam vacuum brake. The Dreadnought vacuum brake is linked through the steam brake valve slightly lower than the vacuum brake. Note the typical North Eastern round-top boiler with the large fire door behind the heat shield. *National Railway Museum, York*

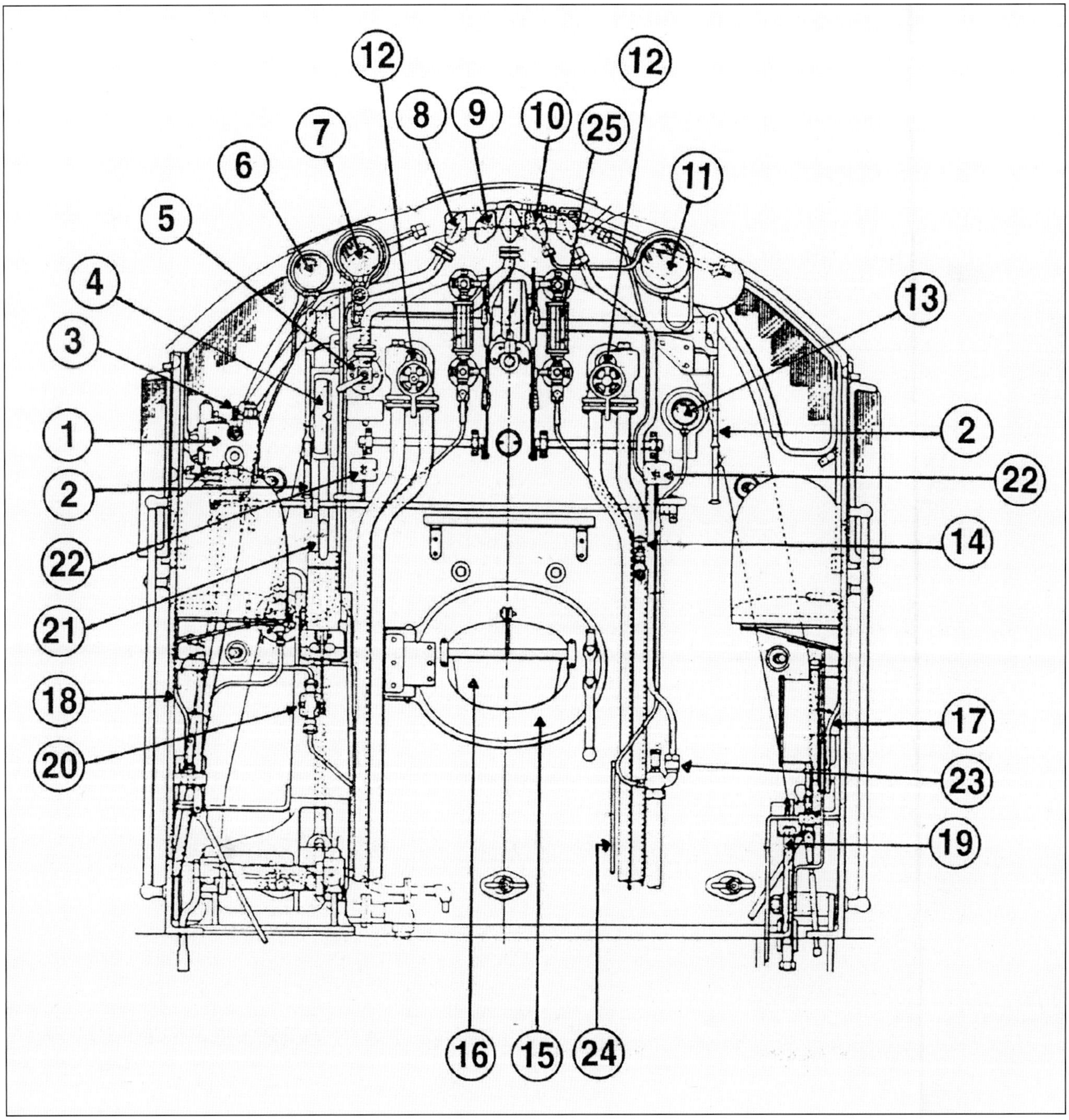

Layout of an LNER A4 class 'Pacific' footplate: 1 vacuum brake assembly; 2 regulator lever; 3 vacuum regulator valve; 4 reversing sector plate; 5 blower valve; 6 duplex vacuum gauge; 7 steamchest pressure gauge; 8 carriage warming stop cock; 9 steam manifold stop cock; 10 pressure gauge stop valve; 11 steam pressure gauge; 12 injector steam spindles; 13 carriage steam heating gauge; 14 coal watering cock; 15 large fire door; 16 fire door flap; 17 cylinder drain cock lever; 18 sanding lever; 19 drop grate screw; 20 reversing lever clutch lock; 21 reversing screw handle; 22 remote control for water gauges; 23 steam heating safety valve; 24 damper rod; 25 water gauges

The footplate of BR Standard Class 4 4-6-0 No 76079. *Brian Dobbs*

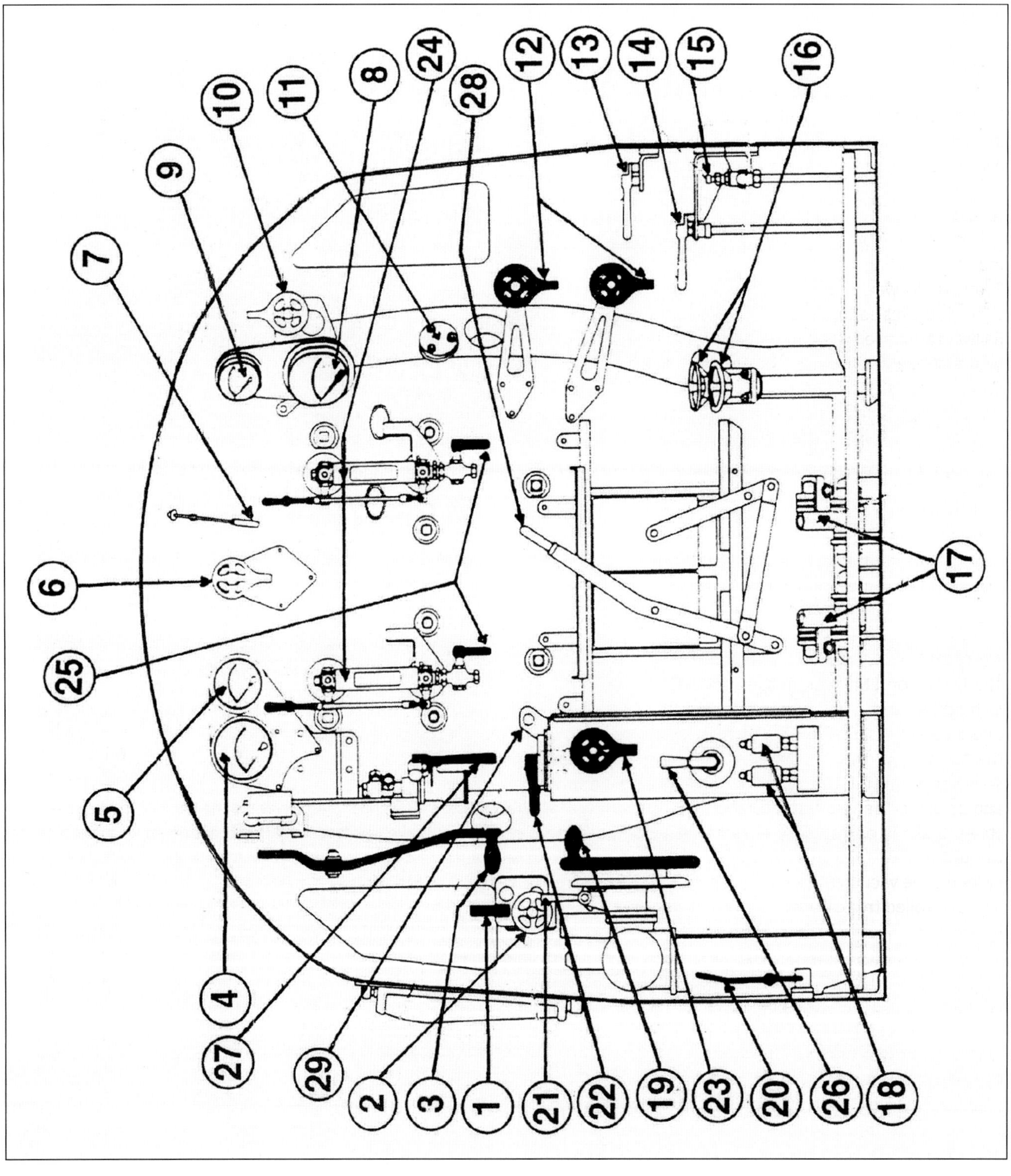

Layout of a typical British Railways Standard footplate:
1 small ejector; 2 large ejector; 3 regulator handle; 4 vacuum gauge; 5 steamchest pressure gauge; 6 steam manifold shut-off; 7 whistle; 8 main pressure gauge; 9 carriage warming pressure gauge; 10 steam shut-off for carriage warming system; 11 blank for continuous blowdown valve; 12 injector steam spindles; 13 live steam injector water handle;
14 exhaust steam water regulator handle; 15 coal watering cock; 16 damper controls; 17 drop grate bar sockets; 18 steam brake oiling points; 19 reversing wheel;
20 cylinder drain cocks;
21 reversing wheel lock catch;
22 driver's vacuum brake lever;
23 blower valve; 24 water gauges; 25 water test cocks;
26 sanding control; 27 driver's graduated steam brake valve;
28 fire door handle; 29 vacuum release valve

BRAKES

The first and most important control with which a driver must become familiar is the brake, which may be steam- or vacuum-operated, or possibly a combination of both, while some locomotives are completely air braked. Before a locomotive is moved, an understanding of the brake and its operation must be acquired.

To move a locomotive is relatively easy, but stopping it where required can take some skill. A driver must always be ready for the unexpected such as the regulator shaft breaking off in the boiler, leaving the valve open. With some locomotives, water pressure caused by priming can jam the regulator open. Such occurrences need instant action on the part of the driver: the brake must be applied and, if possible, the locomotive placed in mid-gear- this will cut off the steam to the cylinders. If the locomotive has been priming, the cylinder drain cocks must also be opened to relieve the water pressure in the cylinders.

In preservation a locomotive's brake controls may be modified to accommodate components from other locomotives, other railways or even other countries. One locomotive that has been modified to bring it up to main-line specification is the former Southern Railway 'Pacific' No 35028 *Clan Line*. It was necessary to fit air brakes so that the locomotive could work with British Railways air-braked stock, making it possibly the only locomotive with a steam brake on the engine only and a vacuum brake on the tender; the latter is used when working vacuum-fitted trains, with the air brake used when working air-braked trains.

The modification started by fitting a Westinghouse steam-driven compressor in the tender, rather than alter the lines of the locomotive by fitting it on the frames. This feeds air for the brakes to the air reservoirs, also on the tender. When working air- and vacuum-braked trains the brakes are applied by a control on a footplate pedestal taken from a Class 47 diesel locomotive and modified to suit *Clan Line*.

Rearview of the tender of No 35028 Clan Line showing the air reservoirs on top and the air compressor fitted in its recess with the exhaust unit on the right. On the buffer beam to the left of the coupling are the two air train pipe hoses, and next to them the vacuum pipe for working normal vacuum-fitted stock. On the other side of the coupling, hanging down, is the steam heating pipe. *R. L. Sewell*

Driver's steam and vacuum combination brake (LMS type)

The brakes on **LMS** locomotives are steam-operated for the locomotive and vacuum-operated on the train. Both are operated by a single lever and built into a common body; the top part of the body houses a piston that controls the steam brake, while attached to the same handle is a valve that admits air to the train pipe to apply the vacuum brake. The steam brake is operated by a shuttle valve in the bottom of the brake body that controls the steam supply to the brake cylinder. The vacuum piston and steam brake shuttle valve are also controlled from a lever and face cam fitted to the driver's brake lever.

When the locomotive is running light engine or working a loose-coupled train (a train without a continuous brake, where the brake operates only on the locomotive and tender), it is only necessary to create vacuum on the locomotive to release or gradually apply the steam brake; the vacuum acts as a return spring for the shuttle valve.

With a light engine or loose coupled train, operation is as follows. When the steam brake is moved to the 'on' position and vacuum has been created in the train pipe by the small ejector, air will be heard to enter the train pipe from the holes in the brake valve. When the brake lever is moved to the 'off' position, the holes will be closed, air will not be able to enter the train pipe and the piston will be drawn in due to the creation of the vacuum. This will cause the lever to be pulled in and the shuttle valve to move and close the steam supply from the boiler to the brake cylinder; the steam brake will then be released. The steam from the brake cylinder exhausts through the shuttle valve to the ashpan.

If the locomotive is coupled to a vacuum-fitted train, the brakes will be released by the use of, first, the small ejector and then the large ejector. When the required 19 to 21 inches of vacuum has been created, the large ejector can be closed and the small ejector should maintain the vacuum against any leaks in the train pipe and brake system.

Driver's graduated steam brake valve (BR Standard types)

The graduated steam brake fitted to the British Railways Standard locomotives does not require the creation of vacuum for its operation. It is worked by steam taken from the boiler and operates on the engine and tender only.

Controlled by a lever on a ratchet-and-pawl mechanism, the further the lever is pulled the harder the application of the brake. When the lever is pulled along its ratchet, a shuttle valve inside the body of the steam brake admits steam to the brake cylinder; when the brake is released the steam is exhausted via the same shuttle valve into the atmosphere.

Driver's vacuum brake valve (BR Standard types)

The driver's vacuum brake valve fitted to the BR Standard locomotives has only two positions, 'on' and 'off'. It works by admitting air to the train pipe and destroying the vacuum, therefore app lying the brake on the train only.

Vacuum-operated steam brake (BR Standard types)

This vacuum-operated steam brake is controlled by a vacuum cylinder situated under the steam brake valve. When the vacuum brake is applied, the loss of vacuum in the train pipe side of the cylinder causes the piston to rise and apply the steam brake on the locomotive.

It is necessary, when easing the couplings, for example, to destroy the vacuum from the reservoir side of the brake. This is achieved by the use of the vacuum release valve on the brake console. If this valve is opened after the vacuum has been destroyed, it will admit air to the system on the reservoir side, which will cause the piston to fall and release the steam brake.

Great Western locomotive brakes

Most Great Western locomotives are fitted with a vacuum brake on the locomotive and tender. The vacuum is created by the ejector and afterwards maintained by the vacuum pump

The controls on the footplate of Clan Line. The large ejector and driver's brake lever, fitted with a clip so that it cannot be used as a brake valve when working air-braked stock, is just below the vacuum release valve. The small ejector, next to the driver's steam brake lever, can be identified by the rag. Just below is the oiling point for the steam brake. Bottom centre is the air brake pedestal from a Class 47 diesel locomotive; on top are the pressure gauges. Other controls visible in the photograph are the sanders and reversing gear. *R. L. Sewell*

The LMS steam and vacuum combination brake controls. At the top is the small ejector valve, and to the left the large ejector wheel. At the bottom is the driver's combined steam and vacuum brake valve, on the face of which can be seen the cam that operates the lever that combines the steam and vacuum brake; on the top of this lever can be seen the hook that holds the steam brake off when the vacuum has been destroyed. Above is the vacuum adjusting valve, sometimes called the 'pepperpot', which is pre-set at the required level of vacuum. *R. G. Fox*

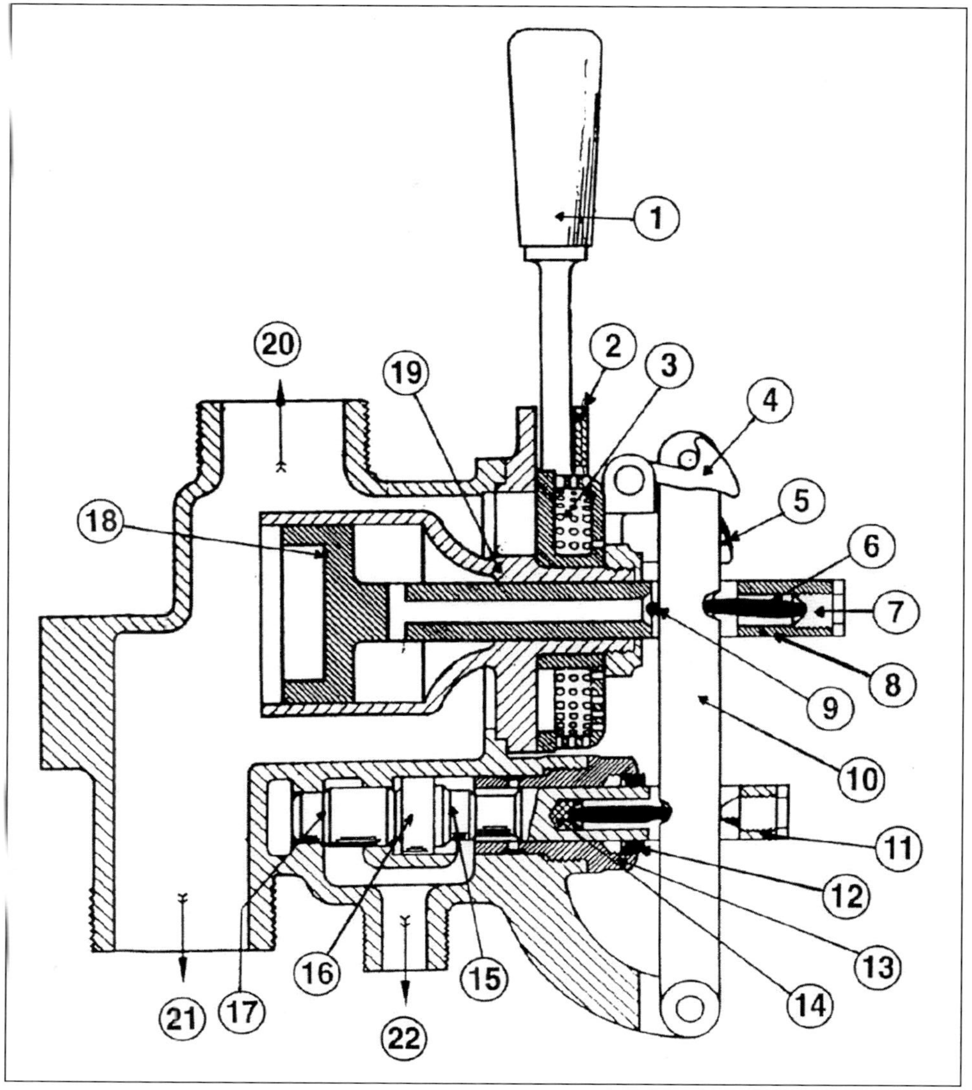

Side-view diagram of the LMS steam and vacuum combination brake controls: 1 driver's brake lever; 2 quadrant plate; 3 disc with air inlet; 4 hook; 5 face cam; 6 pin; 7 cap nut; 8 drilled piston spindle; 9 cross pin; 10 fulcrum lever; 11 steam brake spindle; 12 gland screw; 13 gland; 14 hardened seating; 15 steam exhaust valve and choke; 16 balancing piston; 17 steam brake valve; 18 vacuum piston; 19 disc valve seat; 20 to air relief valve; 21 to train pipe; 22 to steam brake cylinder

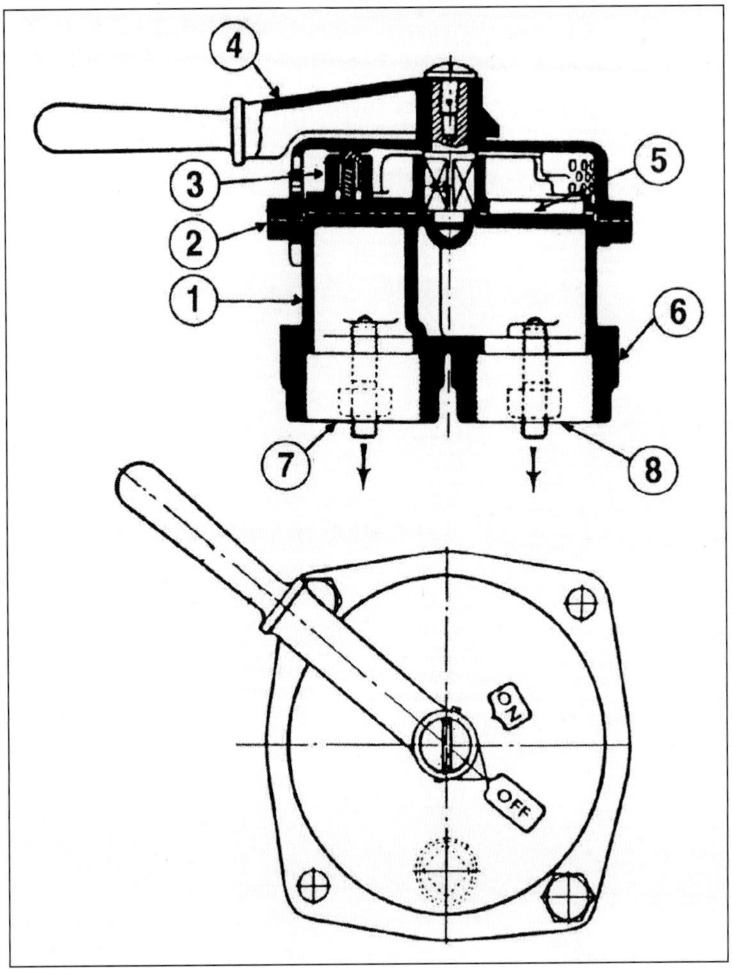

Above: The vacuum brake lever on top of the driver's console on a BR Standard locomotive. As the driver pulls the handle backwards, air is gradually admitted into the train pipe through the holes in the side of the valve. The wheel at the front of the console is the blower valve, and below this is the vacuum release button that destroys the vacuum on the reservoir side of the brake. Just to the left near the reversing gear sector wheel is the small ejector, which is opened by being turned to the left. Further to the left is the large ejector, identified by the hanging jubilee clip, which is operated by pulling the lever. *Brian Dobbs*

Left: Side- and top-view diagrams of the BR Standard vacuum brake application valve: 1 brake body; 2 top cover; 3 plunger; 4 brake handle; 5 disc valve; 6 bottom casting; 7 to ejector; 8 to train pipe

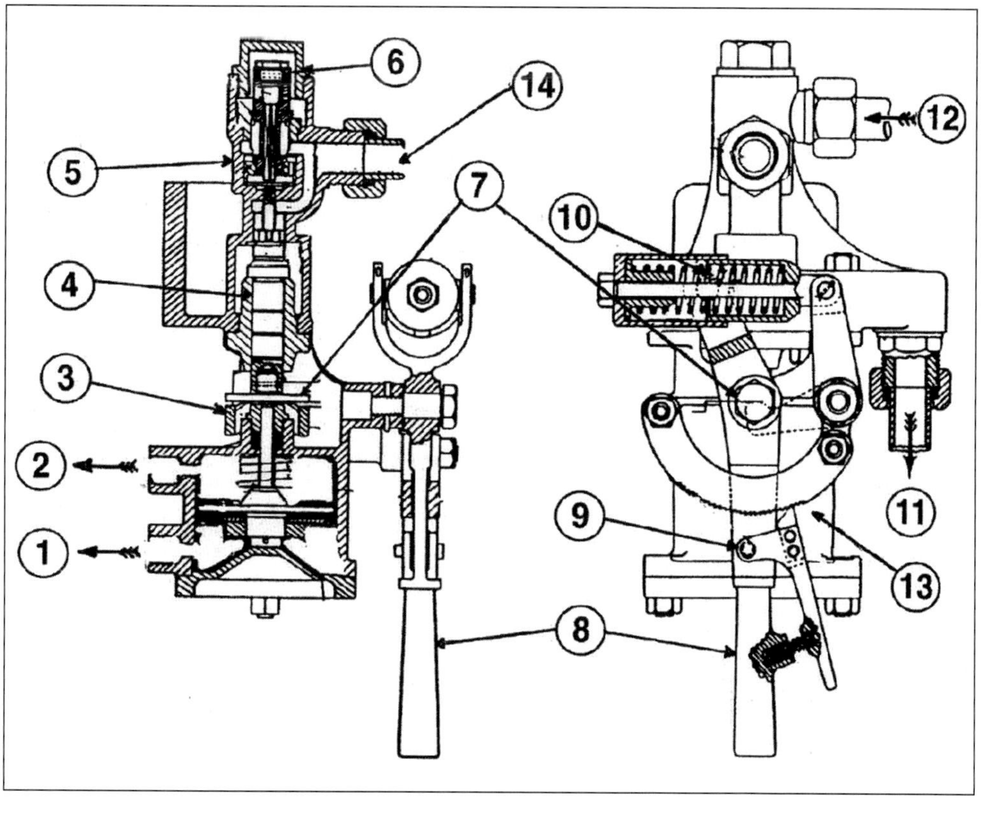

Above: The vacuum-operated steam brake controls on a BR Standard locomotive, identified by the ratchet quadrant. The cylinder that controls the steam brake when working vacuum-fitted trains can be seen with the quadrant bolted to the side. *Brian Dobbs*

Left: Diagrams of the vacuum-operated steam brake controls on a BR Standard locomotive: 1 to train pipe; 2 to vacuum chamber; 3 vacuum brake connection; 4 steam brake bobbin; 5 pilot valve; 6 cap nut; 7 vacuum brake actuating sleeve; 8 steam brake lever; pawl; 10 spring; 11 exhaust from steam brake cylinder; 12 steam from boiler; 13 vacuum control chamber; 14 steam brake valve fitted to the motion of the locomotive. When the locomotive is travelling at a line speed in excess of 15mph the ejector can be closed. The fitting of this pump is to save the steam and water consumed by the small ejector to maintain the vacuum.

Before moving a vacuum-fitted locomotive it will be necessary to create the required level of vacuum. If this is not done it will render the locomotive's brakes non-operative except for the handbrake.

The GWR was unique in adopting a level of 23 to 25 inches of vacuum, whereas other companies adopted 19 to 21 inches for a complete vacuum. It will be noted that the vacuum brake gauge on a Great Western locomotive is divided into two halves, train pipe side and reservoir side. As the vacuum brake was adopted it was necessary to supply a reserve of vacuum, and this was done by fitting a reservoir tank to the locomotive so that the brakes could be released quickly.

The ejector is built into the same body as the brake valve. The valve to the large ejector is on the right-hand side of the vacuum brake assembly just above the blower control. There are only two brake positions, 'on' and 'off'. When the brake lever is to the right the brake is off; it is applied by destroying the vacuum by placing the lever to the left.

There are two methods of uncoupling the engine from a train. The first, with the train stationary, is to completely destroy the vacuum from the train pipe side of the system; this will leave vacuum on the reservoir side and the brakes will be applied on the locomotive and the train. Re-creating the vacuum on the train pipe side, until the vacuum reaches 15 inches, will release the brakes on the locomotive, leaving those on the train still applied. If the ejector is closed so that no more vacuum is created, it will be possible to ease the couplings between the locomotive and train. When the couplings have been eased the vacuum brake is applied by destroying the vacuum again on the train pipe side. The handbrake will now have to be applied otherwise the locomotive will move when the vacuum falls on the reservoir side.

The second method, and the one I personally prefer to use, is to destroy the vacuum completely from the train pipe side. Uncouple the train pipe and place the locomotive pipe on the dummy, leaving that on the train hanging. Now re-create the vacuum so that the locomotive can be moved to ease the couplings, then apply the handbrake before uncoupling. With this method there is no chance of the locomotive accidentally moving the train; also the handbrake is applied with the additional pressure supplied by the vacuum brake.

Should a GWR locomotive need to be towed by another engine, it will be necessary to open the two vacuum release valves on the engine and tender, otherwise vacuum will be created on the reservoir side of the brake by the motion of the crosshead driven vacuum pump, applying the brake.

The two-position brake valve on a Great Western locomotive such as the 'Manor' class. The ejector is built into the body of the brake valve. The lever on the right at the back is the ejector, which is activated by pulling the lever down. *Brian Dobbs*

The three-position steam brake lever fitted to some Great Western locomotives. Just above it is the valve that controls the vacuum brake. When working vacuum-fitted trains and the vacuum is destroyed, the piston applies the steam brake to the locomotive. *Brian Dobbs*

Larger Great Western locomotive brakes

The type of brakes fitted to the larger Great Western locomotives such as the 'Castles', 'Halls' and 'Kings' are similar in design and operation to those of the smaller locomotives such as the 'Manors'. The difference is that in addition to the vacuum pump the locomotives are fitted with an SSJ-type ejector similar to the one fitted to LMS locomotives. This type incorporates a small and large ejector as well as the vacuum pump. The reason for the addition of the small ejector is that when the locomotive is pulling a long train at slow speed the vacuum pump may not be able to maintain the correct level of vacuum against any leaks in the system.

The Great Western steam and vacuum brake

Unlike most Great Western locomotives, the 57XX, 14XX and 45XX classes and some others are fitted with a combined steam and vacuum brake; the steam brake operates on the locomotive only and the vacuum brake on the train. When the locomotive is moving as a light engine or with a loose-coupled train it is not necessary to create a vacuum as with the other Great Western locomotives.

The brake lever has three positions, unlike the two positions of the controls on other previously mentioned GWR locomotives. With the lever over to the right-hand position the vacuum ejector will be operated to create the

Above: The vacuum brake equipment on former Southern Railway No 30777 Sir Lamiel. On the left of the assembly is the brake application lever, which also controls the large ejector. On top of the brake casting is the steam supply valve, while on the right is the small ejector. At the bottom can be seen the vacuum release valve. The two large cap nuts allow access to the steam cones of the small and large ejectors. *Brian Dobbs*

Left: The reservoir tank with the bleed valve at the top left-hand side. *Brian Dobbs*

required 25 inches of vacuum. If the lever is in the upright, or central, position the brake will be off and will not activate the ejector; however, at a line speed in excess of 15mph the vacuum pump should maintain the vacuum with the ejector closed. To apply the brake, the lever is moved to the left; the steam brake will be applied on the locomotive and, if coupled to a vacuum-fitted train, also the vacuum brakes on the train.

There are some important differences with this type of brake. The first is that when running light engine or with a loose-coupled train it will be necessary to leave one of the flexible train pipe connections or bags off the dummy, otherwise the movement of the locomotive will actuate the vacuum pump. This will in turn create vacuum on the reservoir side and apply the steam brake on the locomotive.

When uncoupling from vacuum braked stock it will be necessary to destroy the vacuum in the train pipe, then bleed off the vacuum on the reservoir side by using the bleed valve or jigger valve on top of the reservoir. When the locomotive has eased the couplings, the steam brake is applied with the lever in the right-hand position; therefore it will not be necessary to apply the handbrake as in the case of the other Great Western locomotives.

Before creating vacuum, ensure that the flexible train pipe connections have been replaced on the dummy coupling.

Southern Railway vacuum brake

Like the GWR, the Southern adopted the vacuum brake for some of its locomotives, such as the preserved 4-6-0 No 30777 *Sir Lamiel,* and it operates on the engine and tender. Both the small and large ejectors are built into the body of the brake valve, similar to the design of the Great Western version.

As can be seen in the accompanying photograph, on top of the brake valve is a wheel, which admits or closes off the steam supply to the brake valve. Just to the left is the pre-set vacuum adjuster. The small ejector valve is on the right, and is opened by turning the lever through 90 degrees to the left. The large ejector and the brake application lever are on the left of the brake valve casting. To open the large ejector the lever is pushed fully forward to the stop; when vacuum has been created, the lever is pulled back to the running position, and the vacuum should then be maintained by the small ejector.

The Southern Railway steam vacuum brake valve fitted to 'Pacific' locomotives. On the left, on the front of the casting, is the vacuum brake application lever. On the extreme right of the assembly is the steam brake lever, with its retaining clip. Just above the clip is the vacuum diaphragm that controls the steam brake when the vacuum brake is applied. Above this is the shuttle valve that admits steam from the boiler to the steam brake cylinder. *Author*

The brake application lever has a check valve on the handle controlled by a trigger; this will only partly destroy the vacuum in the train pipe, thus only partly applying the brake. To apply the vacuum brake fully, the application lever must be pulled down gradually, allowing air into the train pipe through the holes in the lever. If the lever is pulled down to the stop it will fully apply the brake. When the lever is returned to the running position, the vacuum will be re created by the small ejector.

Just below the brake assembly casting is the vacuum release valve, which is operated by pulling the small lever to destroy the vacuum in the reservoir.

Southern Railway steam and vacuum brake

The steam and vacuum brake fitted to former Southern Railway locomotives operates with steam on the locomotive and vacuum on the tender. As the vacuum is destroyed the brake is applied on the tender and the steam brake will be applied on the locomotive. To operate the vacuum and steam brake, between 19 and 21 inches of vacuum must first be created by using the small ejector on the front of the brake assembly. As with the previous SR design, the large ejector is built into the vacuum brake application lever and operates in the same way.

The steam brake lever is on the right of the brake assembly and is operated by pulling the lever backwards; this will only apply the brake on the locomotive, not on the tender or on a vacuum-fitted train. Fitted to the steam brake lever is a hook that retains the steam brake in the 'off' position when the locomotive is not creating vacuum. When the locomotive is moved, especially just the engine and tender, care must be taken to ensure that this clip is in the 'up' position, otherwise only the vacuum brake will operate.

At the bottom, connected to the pipe on the reservoir side, is the vacuum release valve; this will drain the vacuum from the reservoir side before the steam brake can be released.

London & North Eastern vacuum brake

The vacuum brake adopted by the LNER 1s controlled by the Dreadnought combined vacuum ejector and brake valve (see also Chapter 1). It is necessary to create between 19 and 21 inches of vacuum before any attempt is made to move the locomotive, otherwise only the handbrake will be available.

The vacuum is created with the large ejector by placing the application brake lever in the 'up' position. When the vacuum has been created the application lever can be placed in the running position and the vacuum will be maintained by the small ejector on the side of the brake valve casting.

Fitted to the application lever is an auxiliary brake valve controlled by a trigger. This is designed to admit air to the brake system by lifting a plug from a socket; when the trigger is released the vacuum will draw the plug back into the socket and the vacuum will rise, releasing the brake.

A complete application of the vacuum brake is made by pulling the lever down.

Air brake (N7 class tanks, etc.)

Unlike most locomotives employing a steam or vacuum brake on the locomotive, the N7 0-6-2 tanks that used to work over the Metropolitan lines in London were required to be fitted with the Westinghouse air brake system in addition to the vacuum brake for the train. As with the vacuum brake, no attempt must be made to move the locomotive without first building up the required level of air pressure, otherwise the locomotive will have no brake except the handbrake.

Air is supplied by a steam-powered compressor fitted at some convenient location on the locomotive. When preparing a locomotive fitted with a Westinghouse air brake, it is necessary to drain and refill the oil supply to the pump (see also Chapter 5 and the diagram in Chapter 1). When this has been done it is important that the air reservoir is also drained of any condensation that might build up from the compressed air. First the air supply in the reservoir is built up using the air pump. When the steam supply valve to the compressor is opened on the footplate, the air pump on the locomotive will be heard to operate and the air will be seen to build up in the reservoir situated in between the frames of the engine. When it has built up to the operating level, the drain valve on the air reservoir, usually under the floorboards on the footplate, is opened and any condensation that has built up will be blown out in the air stream. When the air is clear of water, the valve is closed and the air pressure will be built up again by the compressor to the working level.

If this is not carried out the water will gradually take up the air space in the reservoir, cutting down the amount taken by the brake; eventually the water will find its way into the brake valve.

The N7s were also fitted with a vacuum brake that operates on the train only in a similar method to the steam brake. As with the steam brake, the vacuum is created by a separate ejector with its own steam supply from the boiler.

When the locomotive is connected to and working vacuum-braked stock, the operation is just like the steam brake in so much as when the vacuum brake is applied on the train, the air brake is applied on the locomotive.

The pole reverser. *Author*

REVERSING GEAR/VALVE CUT-OFF CONTROL

The reversing gear for the majority of steam locomotives is manually controlled, although steam-operated reversing mechanisms were successfully used, mainly on the Southern Railway.

The mechanism is usually controlled by one of four methods. The first is a simple pole lever, working in a quadrant and connected to the expansion link by what is called a reach rod. With the lever in the forward position the locomotive will be in forward gear; in the central position it will be in mid-gear, and fully back it will be in reverse. The lever is locked in the required position by a spring-loaded pawl that engages in the slots in the quadrant. When the locking lever at the top of the reverser is pulled, it lifts the pawl from the quadrant to allow the direction of the locomotive to be changed.

Above: The reversing gear or wheel fitted to SR 4-6-0 No 30777 Sir Lamiel. Note the screw thread and the die block working on a pivot connected to the reach rod. The slotted circular plate behind the wheel locks it in the required position. Attached above the screw is the slotted sector plate indicating the direction of the locomotive and the percentage of cut-off *Brian Dobbs*

Right: A close-up view of a sector plate, with the reverser set in mid-gear. *Author*

This type of reversing gear is usually found on shunting locomotives where the change from forward to reverse needs to be quick. However, the cut-off cannot be finely adjusted as with, for example, the reversing wheel because the segments on the quadrant are larger than those on the wheel reversing mechanism.

The wheel reverser operates by means of an in-line thread or worm gear in a cast bracket fitted to the main frame of the locomotive. When the wheel is turned the thread or worm moves a block along the shaft. The block is connected to the reach rod running between the wheel and the expansion link, and is fitted with a pointer that runs along a graduated plate known as a sector plate, which is numbered from 75 (reverse) to O (mid gear) and forward to 75 (full forward gear). The pointer thus both indicates the direction of travel and the percentage of steam cut off for the valves.

On a typical sector plate, such as that fitted to an Ivatt Class 2 locomotive, mid-gear is located between the two number 10s. With the indicator in full forward or reverse gear, steam will be admitted to the cylinder for 75 per cent of the piston's travel, then the valve will close and the steam will expand for the remainder of the piston stroke. As the reversing wheel is turned, the expansion link moves in the motion and cuts down the travel of the valve, causing it to close earlier and allowing the steam to expand more in the cylinder.

The wheel is locked in position by a lever engaging in slots on the wheel. Some screw reversers have a straight bar with two handles, one end of which has a pivoting locking mechanism that controls the pawl that engages in a slotted locking plate behind the bar or wheel.

The third type of reverser is what is sometimes called a 'bacon slicer', and can be found on British Railways Standard locomotives. Its operation is similar to that of the wheel except that the wheel is connected to the reach rod by a gear mechanism to allow it to operate in a forward and backward direction. The cut-off is indicated by a sector drum connected to the reverser. Less common is the reverser on the LNER 'Pacific' locomotives, which has a handle working in a vertical cast iron pillar. The sector plate is a brass slide mounted on the faceplate of the locomotive.

REGULATOR VALVE

Steam is admitted to the main internal steam pipe and the cylinders via the regulator valve, which can be located in the dome or in the smokebox.

The earlier design of regulator valve was the vertical slide type, which was fitted into a large dome on top of the boiler, the size of which had the effect of restricting the diameter of the boiler barrel. This type of regulator valve is controlled by a slide lever handle on the footplate working in a quadrant or bracket on the faceplate of the boiler, and marked 'open' and 'shut'. The regulator handle incorporates a pointer running between the two stops indicating the position of the valve. The handle is connected to the valve by a shaft running through the boiler via a steamtight packing called the regulator gland.

When the regulator valve is opened, steam passes to the main internal steam pipe, either through the superheater and on to the steamchest or direct to the steamchest.

The face of the regulator has a series of ports cast into the fixed body or casting. Mounted on the face is a moving slide,

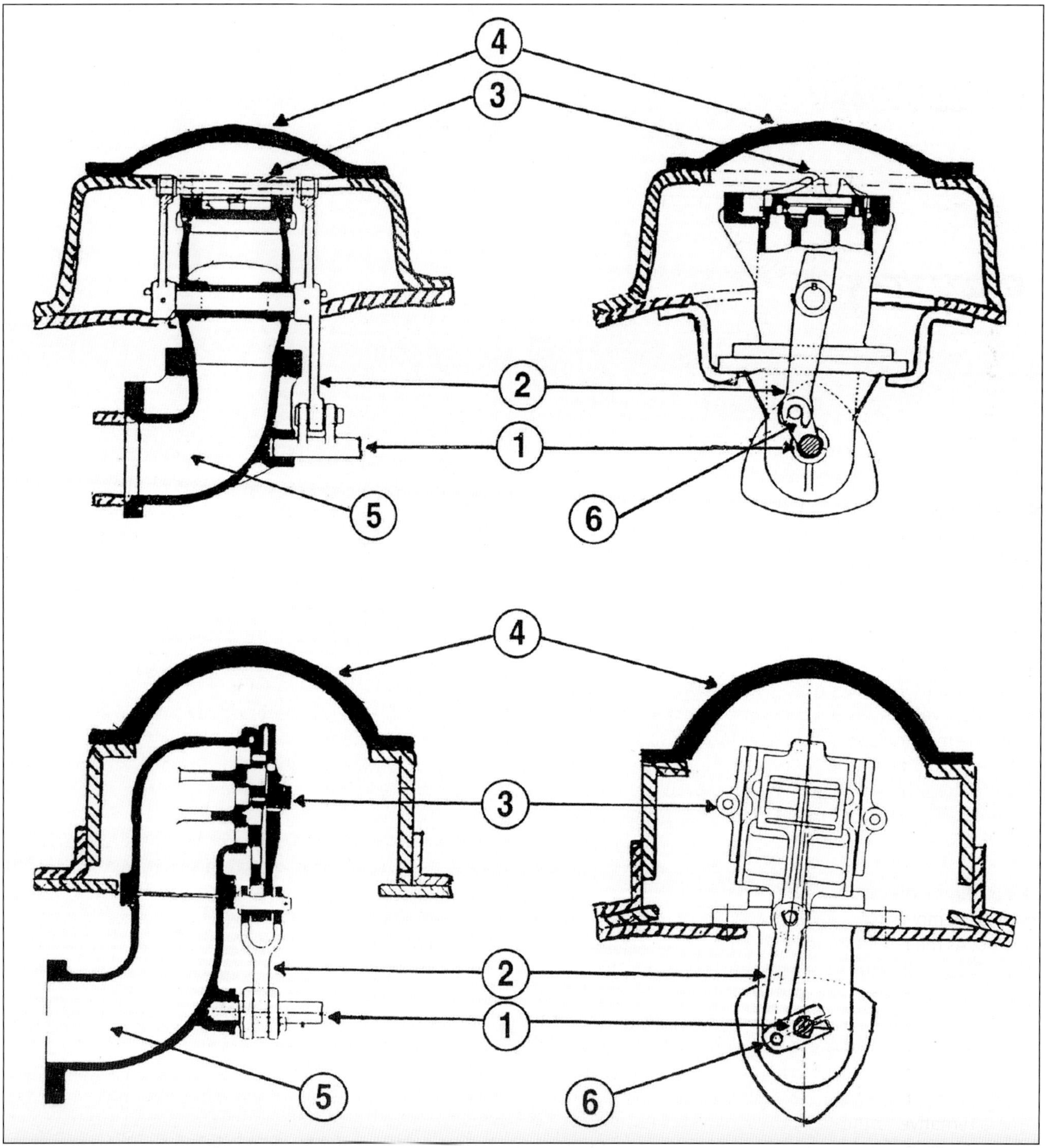

Types of dome-mounted regulator valve: horizontal type and vertical type: 1 regulator shaft with crank, operated by regulator handle; 2 operating lever; 3 slide valve; 4 dome cover; 5 main internal steam pipe; 6 crank

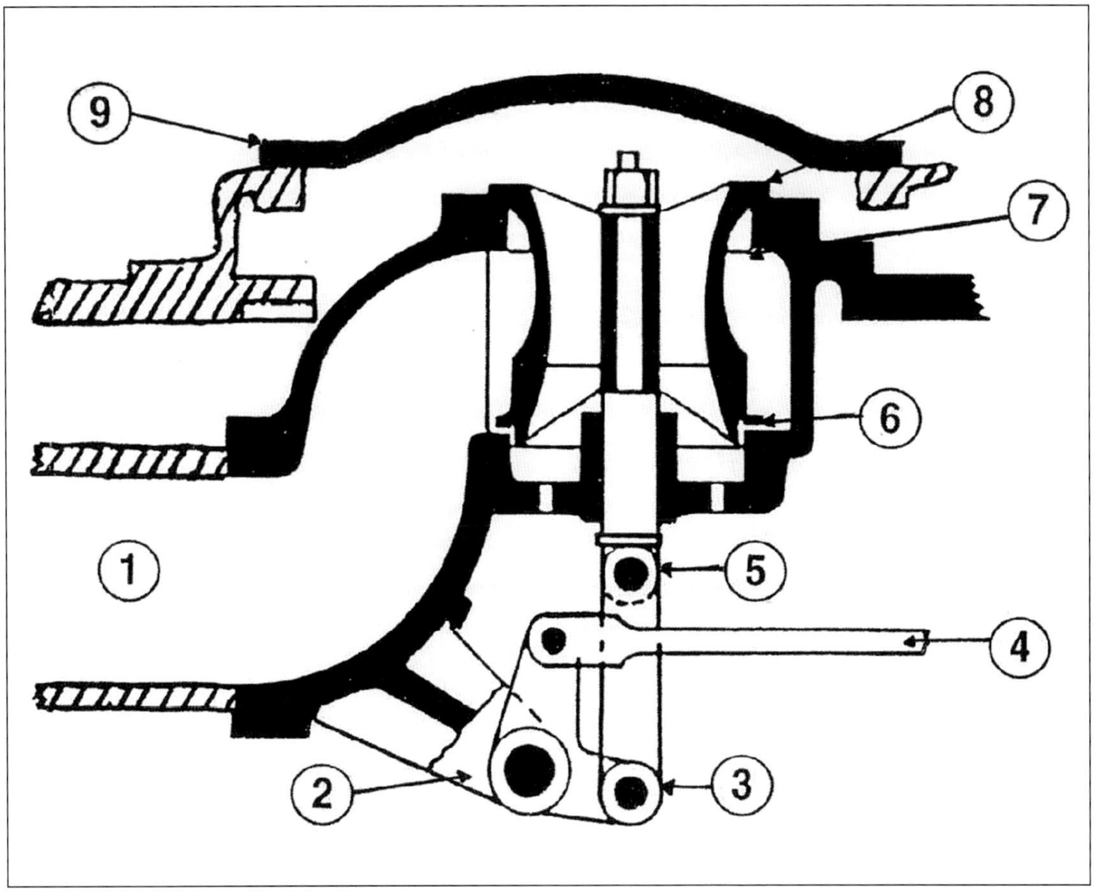

'Double beat'-type regulator valve: 1 main internal steam pipe; 2 bracket; 3 lifting crank; 4 regulator shaft operated by regulator handle; 5 valve push-rod; 6 bottom seat; 7 top seat; 8 regulator valve; 9 dome cover

also fitted with ports. When the regulator is closed the ports of the slide and body are out of line. As the regulator handle or lever is moved on the footplate, the slide moves across the face of the regulator body bringing the ports into line and allowing the steam to pass through all of them to the main internal steam pipe.

The first port to open is the 'pilot' or first valve; when all the ports are brought into line it is called the second or main valve.

The horizontal regulator valve operates in the same way as the vertical version except that the slide moves horizontally across the main casting or regulator body. The reason for this is to reduce the height of the dome to enable the locomotive designer to adopt a larger boiler barrel.

When the regulator lever on the footplate is moved it operates a shaft that is connected to a lever that pivots on the side of the regulator body; this is connected to the sliding part of the regulator valve. When the lever is moved on the footplate, it moves the slide, bringing the ports of the first valve into line. As the handle is moved further, the slide continues to move horizontally across the face of the valve until the second valve position brings all the ports in line.

Another type of horizontal regulator valve is fitted in the smokebox and operates in the same way. It is controlled either by a shaft running through the boiler or by a cross-shaft on the outside of the smoke box.

An alternative to the quadrant regulator lever is the 'pull-out' type, again fitted on a bracket on the footplate with a pointer indicating the position of the regulator valve. The lever is connected by a linkage outside the boiler barrel to a cross-shaft running through a steamtight gland to the regulator valve in the dome or smokebox. This type of regulator is found on later types of locomotive, such as the British Railways Standards.

Another design of regulator valve that can be found in the dome is the 'double beat' type, so-called because the casting employs a double-face valve separated by a bobbin. Unlike the vertical and horizontal valve, there is no 'pilot' and main valve, just a straight valve. The regulator of a locomotive fitted with a double beat valve must be closed gently, otherwise damage will be caused to the seats of the valve.

CYLINDER DRAIN COCKS

Cylinder drain cocks, sometimes called taps, are used to remove any water that builds up in the cylinders by the condensation of steam while the engine is stationary. Another use is in an emergency when the locomotive 'primes', i.e. water is drawn into the cylinders (see also Chapter 4); opening the cocks will prevent damage being caused to the cylinder end covers or the connecting rod.

The cylinder drain cocks are controlled manually via a linkage mechanism from the footplate, usually by the driver although the control is on the fireman's side on certain former LNER locomotives. The BR Standard locomotives have steam-operated cylinder drain cocks controlled by the driver.

When the locomotive is stationary the drain cocks must be left open so that any condensation will drain away, and so that any steam building up in the cylinders due to a passing regulator

The steam sanding control valve in the 'off' position. Moving the handle to the left will operate the front or leading sanders to prevent slipping in the forward direction. Moved to the right, the trailing sanders will operate when the locomotive is in reserve. *Author*

will not cause the locomotive to move. Where a mechanical lubricator is employed, when the drain cocks are in the open position the steam supply to the atomisers will be shut off.

The cocks must also be kept open when a locomotive is moved until the cylinders are warmed up sufficiently to prevent further condensation of steam.

When the locomotive is in motion and priming is detected, the cylinder cocks must be opened and the regulator closed until the water clears. If the priming has been caused by too high a level of water in the boiler, the locomotive can be worked with the cocks open and small regulator openings until the water level drops sufficiently. Likewise, with a greasy boiler the locomotive must be worked with a slightly lower water level to prevent priming.

STEAM SANDERS

The purpose and operation of steam operated sanders has already been described in Chapter 5. The steam supply to the sanders is controlled from the footplate by a three-position duplex valve, directing sand to the front or rear sanders. With the lever in the central position the sanders are off; in the left-hand position the front sanders are operating, in the right hand position the trailing sanders will operate.

If a locomotive starts to slip the regulator must be closed to allow the driving wheels to stop spinning *before* the sanders are operated, otherwise stress could be caused to the motion of the locomotive by the sudden stop.

WHISTLE

The locomotive's whistle is used to warn bystanders of the approach of the locomotive or train, and can also be used to relay messages to the signalman, for example when a locomotive requires water at the station.

Great Western locomotives have two whistles fitted, each with a different tone, one high-pitched, the other lower, the latter for use in emergencies calling for the guard's brake.

7
Driving the train

When the position of driver of a steam-hauled passenger train has been attained, it could be said that one has reached the top of the tree in operating train crew terms. However, the thirst for knowledge of trains and locomotives should never be satisfied - a driver should never feel that he is at the end of the learning curve. Every trip made on the railway will be a learning experience, and no two trips will ever be the same. There are always things to learn and re-learn on the operating railway, increasing the driver's efficiency in the maintenance and handling of locomotives and trains.

A driver on one of today's preserved railways can come into contact with a far more varied selection of locomotives than the professional driver was able to experience on the national railway network in the days of steam. The opportunity for an LNER driver to come into contact with a Great Western locomotive or vice versa was rare, except in the days of the locomotive exchanges that took place just after the Second World War.

It is therefore important that a modern driver should gain as much knowledge as possible, either by reading books on the subject or attending mutual improvement classes that may be held at the railway. This will help both driver and fireman to gain the necessary skill in maintenance, preparation and handling of all the locomotives that he or she may encounter.

THE FOOTPLATE TEAM

All steam locomotives are basically similar in their design, but they can handle and drive in many different ways, with strange peculiarities even among locomotives of the same design. It is true that a driver may go home more tired and fatigued after a day's work than he did as a fireman- not physically but mentally exhausted due to the exacting demands of his duties. Even though the same job, the same locomotive, the same train and the same stretch of line is encountered on each trip, one or more aspects will be in some way different.

The problem may be that the locomotive is steaming badly due to a dirty fire, bad coal, or maybe an inexperienced fireman that he will have to guide and assist. A fireman can look to the driver for guidance or assistance when a problem presents itself, but the driver has no one on the footplate to ask or advise him. Everything rests on his own knowledge and expertise of the subject- the buck stops here!

If the driver has little experience as a fireman he will not be able to guide or instruct his mate if he is in difficulties, so it might become a case of 'the blind leading the blind'. During his time as a fireman the prospective driver should take the opportunity to study the various drivers with whom he works and learn from the experience. Every driver will handle an engine differently, adopting an alternative driving technique for the same class of locomotive. Some may give a poor performance by handling the engine in the wrong way or by being unpredictable in their requirement for steam, for example by not using the same point to shut off steam and allow the train to coast. Others may drive the engine hard, causing it to use excessive amounts of coal and water. With this in mind the fireman should learn the road, for example taking note of where the driver shuts off steam before a station to avoid excessive braking. This will allow him or her to learn the art of driving the correct way; although still demanding, it will become second nature.

While driving a locomotive or train the driver will use his natural senses of sight and hearing: looking for evidence of a hot bearing, a broken part, or an obstruction on the line, in addition to fixed or hand signals, and listening for a loose part or a knocking bearing. Sound is also useful when applying the vacuum brake: the experienced driver will be able to tell by sound the amount of air entering the train pipe via the brake valve without even resorting to looking at the vacuum gauge. Sound is also a guide when the engine is being worked correctly. Even the sense of smell can be helpful in detecting a hot bearing or a rubbing brake.

This chapter will deal with the elementary task of moving a light engine, then progress to a demonstration run with a passenger train. However, first we must consider the all-important subject of safe brake operation.

SAFE BRAKE OPERATION

Before any locomotive or train is put into motion, the person moving it must be fully conversant with the braking system, whether steam or vacuum or a combination of both, or even air-operated, in order to be able to stop the locomotive at the required place or in an emergency. As mentioned in the previous chapter, anyone can move a locomotive, but it takes skill to stop it in the right place. Descriptions of the various braking systems likely to be encountered have already been given in the previous chapter; this chapter is concerned with their effective operation.

The steam brake

This brake is, as its name suggests, powered by steam taken from the boiler. Although at low pressure, eg about l00psi, the brake will still operate, it will not be very efficient; the higher the boiler pressure, the better and keener the brake will be. Steam brakes are only found on locomotives and tenders, never on coaching stock.

BR Standard steam brake

Moving a locomotive fitted with a BR Standard steam brake, there is no need to create a vacuum as the steam brake lever is fitted with a ratchet mechanism that allows a varied application to be maintained without requiring vacuum, just as in the LMS version.

If for any reason vacuum has been created, for example following the working of a vacuum-fitted train, the vacuum in the system will keep the steam brake applied, and will need to be destroyed to release the brake on the engine and tender. This is done by operating the jigger valve or vacuum release valve on the reservoir side of the brake.

LMS combined steam and vacuum brake

This type of brake requires steam from the boiler to operate it; the steam brake works on the locomotive and tender only, while the vacuum side is only for the train. However, it is necessary to create vacuum in the train pipe to control the application of the steam brake, saving the need for the fulcrum lever to be pushed in by hand. There is not usually a reservoir side fitted to this type of brake.

Great Western vacuum brake

Unlike other railway companies, the GWR required 25 inches of mercury for a complete vacuum, and this is created by the ejector before the locomotive is moved. Some of the larger Great Western locomotives have a small ejector, which will be sufficient when moving a light engine. Others have just one ejector, which must be used for both creating the vacuum and maintaining the level when the locomotive is moving so slowly that the crosshead vacuum pump is not able to do so.

With the vacuum brake lever in the 'off' position, open the ejector to create a vacuum on both the reservoir and train pipe sides of the system, then apply the vacuum brake. The additional pressure on the brake rigging by the use of the vacuum brake will make handbrake release easier. With the vacuum brake still applied, select the required direction of travel and, when all is clear, move the locomotive in the normal method.

Great Western steam and vacuum brake

This steam and vacuum brake fitted to smaller types of GWR locomotive works in a similar method to the others. However, there are differences. As with the larger, fully vacuum-braked Great Western locomotives, these steam-braked engines are also fitted with an ejector and crosshead vacuum pump, working in a similar way.

Let us assume that we are about to make a light engine movement. It will be necessary with this type of steam and vacuum brake to remove one of the vacuum pipes from the dummy coupling to prevent vacuum from being built up in the system. The movement of the locomotive will cause the vacuum pump to operate, but this will build up vacuum on the reservoir side only by the action of the crosshead pump. This reservoir-side vacuum will operate a valve in the steam brake body and apply the steam brake.

The ejector on this type of locomotive is built into the brake valve, and controlled by a single common lever with three separate positions. With the lever to the right, the steam brake will be applied; in the central or vertical position the brake is in the running position; to the left it will open the ejector (used when coupled to vacuum-braked stock, not during a light engine movement). Place the steam brake lever to the right to apply the steam brake. The handbrake can now be released and the direction of travel selected before the steam brake is released.

LNER and Southern locomotive brakes

The vacuum brake fitted to these and other locomotives requires the creation of the vacuum before the brake can be applied on the locomotive and train. When such a locomotive is moved it will first be necessary to create the required 19 to 21 inches of vacuum, otherwise the only brake on the locomotive and tender will be the handbrake.

Most LNER locomotives are fitted with the Dreadnought combined ejector and brake lever. When moving a light engine, sufficient vacuum will be created by the use of the small ejector. With the brake application lever in the running position, open the small ejector; this will create vacuum on both sides of the system. When the vacuum has been created, apply the vacuum brake to release the handbrake in a similar way to GWR locomotives. With the vacuum brake applied, select the direction of travel and move the locomotive as normal.

Southern Railway steam and vacuum brake

This brake system is unusual, being in effect two systems. The brake is operated by steam on the locomotive and vacuum on the tender. When the vacuum brake is applied, the loss of vacuum will automatically apply the steam brake on the locomotive. If the locomotive is moved without creating a vacuum, the steam brake will only be working on the engine. It is therefore necessary to create vacuum to supply braking power to the tender, otherwise it will affect the braking distance of the locomotive and tender. Another important point to remember with this type of locomotive is that the clip that holds the steam brake in the 'off' position when the locomotive is standing should not be in the 'down' position, holding in the steam brake spindle, otherwise there will only be the vacuum brake working on the tender. With the vacuum brake in. the running position, create the vacuum by using the small ejector, then destroy it, which will apply both the steam and vacuum brake. As before, this is carried out to release the handbrake. The engine can then be moved in the normal way.

Air brakes

With an air-braked locomotive, the air carries out the work of the steam in the brake cylinder. Therefore, as with the vacuum brake it is necessary to create the air pressure before the locomotive is moved. This is done by starting the air compressor on the locomotive and building up the air supply in the air receiver to the working pressure of 70psi. Open the steam valve on the faceplate; the distinctive tick of the pump will be heard and the air pressure will be seen to rise on the gauge. It becomes second nature to the driver of an air-braked locomotive to listen to the air pump. When the required level has been built up in the reservoir, the bypass valve will open to stop the compressor overfilling the chamber. The air brake can now be app lied, the direction of travel selected, and the handbrake released.

This type of system can be linked with the vacuum brake in a similar method to the steam and vacuum brake of the British Railways Standard and Great Western locomotives. Therefore if vacuum has been created it will have to be released from the reservoir side, otherwise the brake will remain applied.

Note that no description of air braked rolling-stock is given here because of its lack of use on standard gauge operating railways.

DRIVING A LIGHT ENGINE

With the type of brake fitted to the locomotive fully understood, and sufficient steam pressure, vacuum or air pressure as described above, the locomotive can be moved.

First check that the reverser is in mid-gear, the regulator is closed and the cylinder drain cocks are open; all these operations should have been carried out when the locomotive was left or stabled. Apply the locomotive brake and release the handbrake; the additional pressure on the brake rigging will mean that the handbrake will release more easily.

With the steam or vacuum brake still applied on the locomotive, select the direction of travel by placing the reverser fully in the required direction. Check that all is clear and, before the locomotive is moved, sound the whistle.

Release the locomotive brake and gently open the regulator; on some types of locomotive the steam will be heard passing the regulator valve. Continue to move the regulator handle until the locomotive starts to move. Steam will be seen to issue from the cylinder drain cocks as the locomotive gets under way. Allow it to move slowly for about six or eight revolutions of the wheels to clear any condensation that has built up in the cylinders and superheaters. If it has been standing for some time, the cylinders will be cold and may take a few more revolutions to clear any condensation. Droplets of water from the cylinders will be seen at the chimney top, and as soon as the water clears the cocks can be closed. With a mechanical

As the train starts to build up to the required speed the reversing wheel can be slowly wound back from full gear, reducing the cut-off until the locomotive is working economically. *R. G. Fox*

lubricator, when the cylinder drain cocks are open the steam supply to the atomisers is closed off, so no lubrication will be fed to the cylinders; it is therefore beneficial to close the cocks as soon as possible.

During this time a vast amount of steam will be circulating around the locomotive from the drain cocks, causing poor visibility, especially when moving chimney-first, so it will be necessary to be vigilant and if necessary sound the engine whistle.

Using the reversing gear/ valve cut-off control

Wheel reversers, as found on larger locomotives (see also Chapter 6), provide a fine percentage of cut-off of the steam following the piston. Some drivers will only use 50 per cent cut off or less on the reversing wheel or lever as they move away to prevent the locomotive from slipping, but it is recommended that full valve travel is used.

As the locomotive reaches a reasonable speed the reverser should be 'wound up' two full turns and the locomotive allowed to accelerate. As the speed rises the reverser can be wound up further; care must be taken not to choke the locomotive by leaving too much steam in the cylinders.

As already described, when a locomotive is in full gear steam enters the cylinder and forces the piston along for 75 per cent of its stroke, then the valve closes and the steam expands following the piston. As the piston reaches the end of the stroke it reverses direction and exhausts the steam for 75 per cent of the stroke, then the valve closes, leaving steam in the cylinder. This is then compressed to form a cushion of steam just before the end of the piston's stroke. The valve then opens to admit lead steam. If too much steam is being used it will be wasted by not being allowed to expand in the cylinder, therefore using excessive amounts of coal and water.

The additional blast will cause the fire to be thrown up the chimney.

If the reversing gear is the older pole-type lever, there will not be the same fine control of the valves as in the screw type because the valve setting is controlled by the slotted quadrant plate, with only a few segments. An equal number of segments are supplied for forward and reverse direction of travel, with a central position for mid-gear. With this type of gear, when adjustment of the cut-off is required the regulator may have to be closed momentarily to enable the lever to be moved against the steam pressure in the steamchest. This is where experience of the line and the locomotive is valuable, enabling the driver to estimate the required position for the cut-off. With the regulator closed, the lever is pulled back and secured in the required position by the locking catch. The regulator is then re opened and a check made that the required setting has been achieved; if not, the operation must be repeated until the locomotive is working efficiently.

Using the regulator

Both vertical and horizontal slide regulator valves operate in the same way. As the regulator is partly opened resistance will be felt, indicating that first or pilot valve is open. Usually this will be sufficient for light engine movements. Further movement of the regulator lever past the resistance will open the second or main valve. To do this it may be necessary to place the locomotive in mid-gear to overcome the resistance of the steam in the regulator valve.

When the train is under way and moving at a speed that will allow time to do so, the regulator valve can be closed and the reverser moved to almost mid-gear; then the regulator lever can be opened fully and the reverser can be let out slowly in the direction of travel. The speed will be felt to increase as more steam is admitted to the cylinders. When the required speed has been reached the cut-off can be set to give an economical performance.

With the pole-type reverser the regulator will usually have to be closed before selecting the cut-off. When the train is under way the regulator is closed and the pole moved to the required position. With the cut off reduced it will usually be possible to move the regulator into main or second valve. The regulator may also have to be closed before any adjustment of the cut-off is attempted. When moving forwards with the regulator open, if the catch should be accidentally released the lever would fly forward violently into full travel.

When closing the regulator it is necessary to make sure that the pointer has closed properly against the stop. It is easier to close the regulator quickly as this will cause the slide to shut off the supply of steam.

Stopping

A locomotive is stopped in the following manner. First close the regulator and allow the engine to coast. If the reversing gear has been wound up, it must be returned to full gear and the steam or vacuum brake applied, much in the same way as a motor car. If the brake is applied all at once the locomotive will stop with a jolt, and if the rail is wet this will cause the engine to slide; this is referred to as 'picking its wheels up'. Therefore it is advisable to shut off steam early, allow the engine to run and brake early, rather than leaving it to the last moment. Indeed, it is good practice to allow the engine to coast to a stand with only a little use of the brake; this will cut down to a minimum wear on the tyres and brake blocks of the engine and tender.

When the locomotive is stationary, and with the steam or vacuum brake on, the handbrake must be applied. The additional pressure of the steam or vacuum on the rigging will make the application firmer. Once the handbrake is applied the steam or vacuum brake must be placed in the 'off' position and the ejector closed. On those locomotives fitted with the combined steam and vacuum brake, such as the LMS type, the brake must be pinned up by hooking the steam spindle on the clip. It is not good practice to leave the locomotive standing with the steam brake applied as this will blow the oil out of the brake cylinder, causing rapid wear and the piston to stick in the cylinder. The locomotive must now be placed in mid-gear and the cylinder cocks opened to stop the build-up of steam in the cylinders.

BASIC FIXED SIGNALS

The standard semaphore signals are the red and white stop signal and the yellow and black fish-tailed distant signal, the latter warning of the aspect of stop signals ahead. The stop

Above: Two stop signals controlling the exit from a station. The one on the left is in the 'on' or danger position, while the one on the right is in the 'off' or clear position. The white diamonds on the signal posts denote that the line is 'track circuited', i.e. the presence of a train will be indicated in the signal box by lights on the signalman's track diagram. *Author*

Left: A distant or caution signal warns of the aspect of the stop signals in the section ahead. Where distances between signal boxes are short, the distant arm for the section ahead may be mounted beneath the starting signal of the previous section, as here. The distant arm can only show clear when the stop arm above it also shows clear, and both arms return to danger/caution together. *Author*

A bracketed stop signal controlling a junction. As can be seen the posts are of different heights, the taller one on the left indicating the straight route and the lower one indicating the divergence to the right. Below each arm is a smaller 'calling on' allowing a locomotive to pass the main arm at danger and enter the line ahead which is already occupied by a train, for shunting purposes, etc. *Author*

Beneath the stop signal can be seen a subsidiary ground or disc signal controlling a shunting movement into the siding. The disc is in the 'off' or clear position and the points set accordingly; signal and points are mechanically interlocked to prevent conflicting aspects. *Author*

A pair of ground signals; the upper disc applies to the route straight ahead and the lower one to the diverging line. *Author*

Trap points, facing the direction of travel, are designed to protect a crossing or a line running in the other direction; should a locomotive or train pass the signal at danger it will be derailed rather than allowed to run out on to the main line. When it is safe for a train to leave, the points are pulled just as with standard points. Catch points, facing away from the direction of travel (or 'trailing'), are found on the up grades of inclines. They are similar to trap points but are spring-loaded and not controlled by a signal box or ground frame. They are designed to catch any runaway wagons travelling in the wrong direction along a running line and derail them. *Author*

signals are described as 'home' or 'starting' signals (in more complex layouts there may be additional stop signals known as 'outer homes' and 'advanced starters'). The home signal is before the signal box controlling the section or before a junction, while the starting signal is beyond the signal box and controls the passage of trains into the next section. The section of line controlled by one signal box, and perhaps including a station and goods yard, is called a 'block section'. Semaphore signals exhibit two positions: clear or 'off', when the arm will be up (upper quadrant) or down (lower quadrant), and danger or 'on' when the arm is horizontal. The distant signal will only show a clear aspect when both the home and starting signals of the section ahead are 'off'. The distant may be on its own post or beneath one of the stop signals of the previous section.

There are several types of smaller subsidiary semaphore arms that may be found on the same post below the home signal; one is known as a 'calling on' arm, and is used for shunting movements. It indicates that the line is occupied by another train or engine, but when cleared allows the driver to proceed at caution to the point of obstruction or to the next stop signal. Where there are two or more home signals controlling a junction they will often be mounted on posts of different heights; the highest signal indicates the straight or principal route, while the lowe one(s) indicates the divergences.

Ground signals, usually situated as their name suggests at ground level, are used for shunting movements within a station or goods yard limits. They are generally a white disc with a red stripe (sometimes a miniature arm), and there may be as many as four or five on top of one another depending on the complexity of the junction, but they always read the same: the top disc indicates the furthest left divergence, and so on. Another type of disc ground signal has a yellow stripe on a white background; this is just a direction signal and can be passed at danger unless the required diversion is necessary.

Before a driver takes a train out on the line he must be fully conversant with all these signals and their operation. He must also be conversant with **trap points.** These are designed to protect a section of line or junction from being entered or fouled by a train or engine where permission has not been granted for the movement.

Hand-operated points are often found in goods yards and depot yards. They may be right-handed or left handed: looking in the direction of travel, if the continuous or 'stock' rail is to the right it is a left-handed point, and if the stock rail is on the left it is right-handed.

When the points are pulled it is important that the blades move over fully and are not standing off, otherwise the points may split under the wheels, causing the locomotive or train to become derailed.

A JOURNEY WITH A FORMERLMS LOCOMOTIVE

As we did as a fireman in Chapter 4, let us now take a journey on a preserved railway with the same type of locomotive and train to help us understand the duties and actions of the driver. The locomotive will be the same, a 'Black Five', because as mentioned before it has many aspects in common with other engines used on operational private railways. However, I will deviate on occasions to cover some of the differences with other locomotives. As with the fireman's trip, the journey will be only an example, as so many railways have their own rule books and differences in signalling, etc.

Leaving the depot

We have arrived at the depot on or before time and the locomotive has been prepared; all the duties have been completed, and the engine is ready to leave. It will be second nature for the driver, as it was for the fireman, to check the level of water in the boiler and tank. The fireman will be attending to his fire as the driver makes a visual check of the locomotive tools and oil. Let us assume that the preparation crew has done a good job and all is in order, and we can now leave the depot to collect the coaching stock.

The 'Black Five' is fitted with an LMS steam vacuum brake. This should be operated as already described: vacuum created with the small ejector, steam brake applied, vacuum destroyed, and handbrake released. If all is clear the whistle is sounded and the locomotive is moved, again as previously described. On our way to collect the coaching stock the cylinders will warm up and the drain cocks can be closed; this will open the steam supply to the atomisers.

Coupling up

As the locomotive approaches the coaching stock it is usual to be called back on to the train by hand signals from the shunter or guard.

If the rail is wet or greasy, as the locomotive approaches the stock it is advisable to open the rear sanders in good time to place a bed of sand on the rail to prevent the locomotive from slipping when starting the train. When the sanders are placed on, it will take a few seconds to clear the water from the system before they start to operate. The sand will also assist the braking of the locomotive, preventing it from colliding with the train. A point to remember when approaching coaching stock either tender-first or chimney-first is that there is a blind spot varying from a minimum of about 3 feet tender-first to about 6 or 8 feet chimney leading; the need to be called on to the train can thus be easily appreciated.

When the locomotive has buffered to the train the regulator is closed, the handbrake applied, the reverser placed in mid-gear and the cylinder cocks opened. Assuming that we are going straight from the sidings to the station, the fireman will couple the engine as already described (Chapter 4). It is always beneficial if the responsibility for applying and releasing the handbrake should rest with the fireman; this increases his safety while he is down between the engine and train.

When the guard has completed his duties, he should inform the driver of the weight and composition of the train.

Creating the vacuum

When the fireman returns to the footplate the driver can create the vacuum, first by opening the small ejector, then the large ejector. When the required level of vacuum has been achieved, the large ejector can be closed and the vacuum will be maintained by the small ejector. It is important that the large ejector is closed when the vacuum has been created, otherwise damage can be caused to the vacuum chambers on the train.

If the small ejector will not maintain the level of vacuum, look for possible leaks at the joints in the train pipe; these can be detected by the sound of air entering the system. It must be assumed that the problem lies with the train, as the vacuum level was satisfactory when the locomotive was prepared. Perhaps while coupling up one of the rubber sealing washers has become displaced or lost. It is also possible that the communication cord has been pulled by vandals or someone leaving the carriage- look along the coaches for the turn button situated in the top corner; one in the horizontal position indicates the coach where the cord has been pulled. Another possible fault is that the steam pressure is low or insufficient to operate the ejector; every locomotive has a minimum pressure at which the ejector will create vacuum.

After the vacuum has been successfully created, and before the train moves away, the guard will carry out a vacuum brake test by opening the brake handle m his brake compartment, admitting air to the train pipe; this will be noted by the needle on the footplate vacuum gauge moving to near zero. When the guard releases his jigger valve the vacuum will rise again. If this test was not carried out and the vacuum had for some reason been drained from the train - if, for example, the stock had been shunted to make up a train with other stock and the pipes had not been properly re-connected - it might mean that only the front part of the train had an operational continuous brake. The brake test will prove that the continuous vacuum brake required by Department of Transport regulations is operating on the full length of the train and that there is no obstruction or break in the train pipe. If there was an obstruction, again the vacuum would only be created in part of the system.

After the test the guard will instruct the driver to recreate the vacuum with the large ejector; again when this has

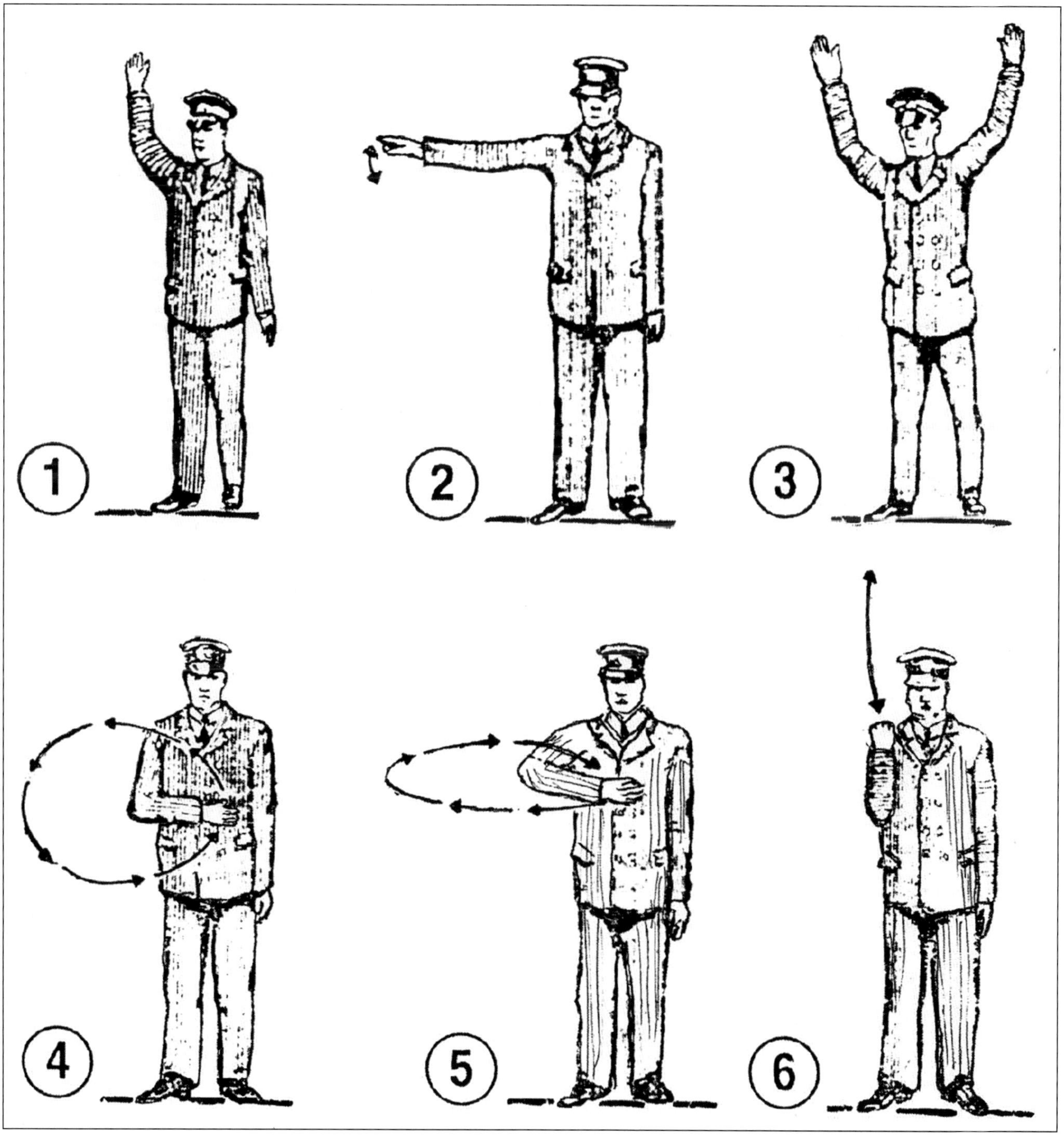

Hand signals: 1 an arm held above the head denotes 'all right'; 2 an arm held out horizontally with the hand moving up and down denotes 'caution' or 'slowdown'; 3 both arms held up denotes 'danger' or 'stop'; 4 an arm moved in a circular manner away from the body denotes 'move away from the hand signal'; 5 an arm moved across and towards the body at shoulder level denotes 'move towards the hand signal'; 6 an arm moved up and down above the shoulder denotes 'create vacuum'

been done the large ejector must be closed. Another point to remember when collecting stock is that, if the train was shunted or worked in with a Great Western locomotive operating on 25 inches of vacuum, the cords fitted to the brake cylinders will need to be pulled to equalise the vacuum level in the brake system. This will have to be carried out with no vacuum being created by the locomotive. If it was not done the vacuum brake would remain partly applied. If it is suspected that the train has been connected to a locomotive operating on 25 inches of vacuum, it is good practice after the brake test for the driver to check that the pistons of the train cylinders have fallen to the bottom of the chambers. There is, of course, no need to check this if the train is being coupled up to another Great Western locomotive creating 25 inches.

Moving to the station

When the time comes to take the train into the station, the guard will give the signal to start. The locomotive is moved in the same way as was described for leaving the shed; if it has been standing it will be necessary to leave the drain cocks open for a few revolutions of the wheels to clear any condensation that has built up in the cylinders.

With the additional weight of the coaches, a greater opening of the regulator will be necessary to move the train away from a stand. Care should be taken not to open the regulator so much as to make the engine slip; this will come with experience and by listening to the locomotive. If a slip occurs, close the regulator and allow the wheels to stop spinning before opening the sanders; as already mentioned, do not use the sanders when the wheels are slipping as this will cause a sudden shock to the motion that could bend a coupling or connecting rod, or damage a crank pin. Once the sanders are on, try again with gentle repeated openings of the regulator for a short time until the train is under way. Should the locomotive slip while the sanders are operating, close the regulator and wait until the slip ceases, then slowly re-open the regulator.

On the way to the station the fireman should be carrying out his duties by preparing his fire for the journey. Also on the way to or upon arrival at the station, the vacuum brake should be tested for its efficiency. When the train reaches the station the vacuum can be destroyed and the handbrake applied on the locomotive; once more the cylinder drain cocks should be opened, shutting off the steam supply to the atomisers and thus saving lubricating oil; it will also prevent condensation from building up in the cylinders. The locomotive must also be placed in mid-gear. When the train is secured, the steam brake must be pinned up in the 'off' position.

A point to mention here is that, if the locomotive is fitted with the British Railways graduated steam and vacuum brake and certain others, the jigger will have to be used to drain the vacuum from the reservoir side, otherwise the steam brake will still be applied.

Right away!

As departure time approaches, the signals will clear to the 'off' position. Select the direction of travel and create the vacuum as before to release the vacuum brake from the train ready for departure. Just before the booked time, the handbrake can be released, the train being held stationary with a complete vacuum by manually holding out the steam brake spindle to keep the steam brake applied; this can be done by jamming the steam spindle out with the retaining hook in the vertical position, or by placing a piece of wood between the central lever and the body of the brake valve.

On receiving the 'right away' signal (sometimes called the 'train ready to start' signal) from the guard, check that the signals are still clear, and knock the steam brake handle to release the hook or remove the piece of wood; this will allow the steam brake spindle to return and the brake will be released. Open the regulator and start to move the train from the station. As before, if the locomotive has been standing the drain cocks should be left open for one or two revolutions of the wheels. After about one or two revolutions close the regulator, then re-open it a little more and for a little longer as the train accelerates; by now the drain cocks should have been closed.

On the road

Here different methods of driving will become evident, the use of the regulator being the cause of a certain amount of controversy. One method is to use first valve all the way, then to leave the locomotive in first valve and control the speed by the use of long valve cut-off; this is usually desirable with freight locomotives that are mainly used for slow-running trains. As soon as the line speed has been reached, the cut-off can be decreased by winding up the reversing gear by about 10 per cent, then the locomotive allowed to settle for about a hundred yards, repeating the process until the required setting has been achieved. The locomotive exhaust should sound even with a crisp 'bark'; there should be no sensation of the locomotive oscillating, caused by winding the locomotive too far back.

Another method is to use second valve all the time and a short valve cutoff; this is more suitable for larger express locomotives designed for fast running, with maximum expansion of steam in the cylinders and long lead steam.

The train will have been started in pilot or first valve to prevent slipping. Once the line speed has been attained, place the locomotive nearly in mid gear, then place the regulator right over to second valve. The reverser can then be used to adjust the valve travel by winding it out in the direction of travel to maintain line speed. Driving a locomotive in this way allows greater expansion of the steam in the cylinders. Care must be taken not to wind up the valve travel too soon, thereby strangling the progress of the locomotive.

Whichever method is used, during the early part of the journey the locomotive must be worked lightly so as not to blow holes in the firebed, making the fireman's job difficult. The firebox and brick arch will still be relatively cold, making the locomotive more difficult to steam. Also, with a comparatively cool boiler the locomotive will have a tendency to prime; as already described, should this happen the cylinder drain cocks must be opened and the regulator closed until the water clears. The cocks should be closed again as soon as possible and the locomotive worked lightly until priming ceases.

Conversely, if a locomotive is worked so lightly that the fire does not receive enough draught through the firebed, it may be difficult to maintain steam pressure. Again, if the reversing gear is wound back too much the locomotive will start to 'hunt' and a slight oscillation on the footplate will be experienced.

If the locomotive is driven with too much steam, with large regulator openings in conjunction with a long valve cut-off, the intense blast will cause excessive amounts of fuel to be burned. This can be detected by a loud blast at the chimney top.

The suitability of these driving methods will also vary with the type of locomotive, the weight of the train and the gradient of the line. It is reasonable to assume that at some time during the journey the loco motive will experience a down grade where steam is not required. To coast with an LMS locomotive with Walschaerts valve gear, for example, the reverser should be placed in the 'drift' position, marked 'D' on the sector plate, and the regulator should be opened slightly. This will ensure that a breath of steam enters the cylinders to form a cushion for the pistons and prevent the valve gear from moving unchecked; it will also supply lubricant to the cylinders. Coasting with the regulator just open prevents ash from being drawn by vacuum from the smokebox into the blastpipe and finding its way into the valve chest and cylinders. It will also save wear on the anti-vacuum valves, keeping them on their seats and leaving their use for the last few yards should the train be coming to a stop.

During the journey the driver must keep a look out for any obstruction or signals along the route and act accordingly. Tunnels can present a hazard, causing blow-backs of fire on to the footplate from the fire doors, so it is advisable to open the blower valve a little wider before entering a tunnel; a similar effect can occur when passing beneath an overbridge or through a deep cutting. When passing a signal box the driver and fireman must look for any hand signals that the signalman may give.

Braking and stopping

We will be soon arriving at our first station stop. Before the regulator is closed the driver should check that the blower valve is open, to prevent a blow-back of fire on the footplate. To stop the train, first the regulator is closed, making sure that the pointer is fully against the stop, then the train is allowed to coast with the reverser placed in full gear. Sometimes this causes the reversing wheel to rattle; this can often be alleviated by winding it back half a turn and locking it.

Stopping a passenger train in the right place without causing oscillation in the couplings is an art that can only come with practice. Most drivers will have a shut-off point and a place where they will start to apply the vacuum brake. It is desirable to bring the train to a stand with the vacuum rising in the system, so that the brake will be releasing on the front portion of the train. It is important that, on the approach to a station, the train has sufficient momentum to apply the brake gradually; if it is moving too slowly it will stop the instant that the brake is applied. Conversely, if it is going too fast hard braking will be required, causing discomfort for the passengers.

Thus on the approach to the braking point before a station the vacuum should be destroyed to around 12 inches in the train pipe, until the braking is just felt. At that point the vacuum brake lever can be returned to the running position to re create the vacuum and allow the train to coast; the application can be repeated if necessary to bring the train under control. This will open all the direct admission valves on the train pipe, causing air to be admitted to the brake cylinders; the pistons will rise in the chambers and apply the brakes.

These direct admission valves are an important part of the continuous braking system, ensuring that the train stops smoothly. If the vacuum brake operated as on freight stock, it would be difficult to stop the train smoothly as the braking effort would start at the front of the train. With the direct admission valve the brake is applied on all the vehicles equally along the whole length of the train, the air being admitted to the vacuum chambers via the direct admission valves on the train pipe rather than via the brake valve on the locomotive.

When the brake is felt to be slowing the train, the brake lever is put back into the running position to allow the ejector to re-create the vacuum and release the brake. This will cause the vacuum to be re-created first on the leading coach, closing the direct admission valve. As the vacuum is re created the brakes on the leading coach will be released, and the vacuum will then rise along the train; as this happens it will close the direct admission valves on the subsequent coaches, releasing the brakes in turn from the leading portion as it returns to the correct level, and taking the braking effect to the rear portion of the train.

If the brake application has been too great, it can be released quickly by the use of the large ejector; as soon as the brake is released, the large ejector should be closed. It is bad practice to open the regulator again, as it will cause discomfort for the passengers in the train.

Station stop

When the train is at a stand in the station the vacuum brake lever should be put in the running position and the vacuum re-created, releasing the brakes on the train. It is, however, important that the train should not move while at the platform; even a slight movement could cause injury to an alighting passenger. The train is therefore kept stationary as described earlier, by holding out the steam spindle to keep the steam brake applied on the locomotive. A locomotive fitted with a British Railways graduated steam and vacuum brake can be kept stationary by applying the steam brake and placing the vacuum brake in the running position.

This first stop provides a good opportunity to check the water level in the boiler and the general performance of the fireman. If the water level is considered to be low, it is better to wait at the station until it is practical to re-start the train. To start in a bad position, for example low steam or low water in the boiler, could mean that the vacuum might start to fall due to the lack of steam. Should this happen it is second nature to open the large ejector to re-create the vacuum, but this may not work due to the low steam pressure; indeed, in some cases the use of the large ejector will cause the level of vacuum to fall more quickly. It would be a bad practice to have to stop the train due to lack of steam in a tunnel, for example, as the confined space would cause choking conditions on the footplate.

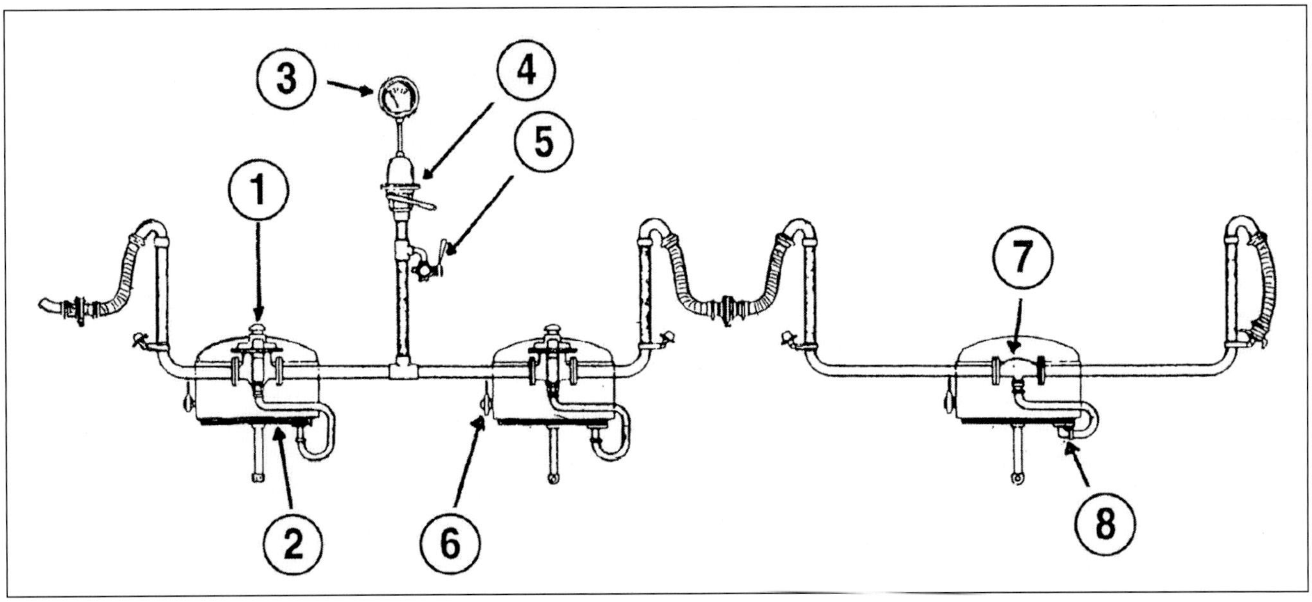

The train pipe and brake cylinder arrangement installed beneath passenger (left) and goods (right) rolling-stock. The vacuum brake gauge and setter are fitted in the passenger guard's compartment only. Not shown is the connection to the communication cord equipment: 1 direct admission valve; 2 brake cylinder; 3 vacuum gauge; 4 brake setter; 5 1-inch cock; 6 release valve; 7 'T' piece; 8 moderating elbow

If a member of the public wishes to visit the footplate during the stop they must be warned that everything is hot to the touch. Visiting children should be accompanied by a responsible adult and a watch must be kept on them so that they do not expose themselves to danger. It is advisable to close the firebox doors.

Emergencies

When the 'right away' has been received the steam brake is released and the train started just as before. Let us imagine, however, that line speed has been achieved when suddenly the vacuum starts to drop to about 12 inches - it can be assumed that the communication cord has been pulled by one of the passengers. The driver must make a decision whether to bring the train to a halt immediately and check on the problem, or proceed to the next station. On a preserved railway it will usually be easiest to stop at the next station unless the cause can be seen; the effect of the brake application can be over-ridden by the use of the large ejector. In any event, the train must not be stopped on a viaduct or in a tunnel because of the danger to passengers.

If, on looking back, the driver notices that the train is on fire he must, if it is safe to do so, bring the train to a halt as the passing wind will fan the flames. Again, the train must not be stopped on a viaduct or over bridge or in a tunnel. When it has been brought to a halt the passengers will have to be de-trained or moved to a place of safety. The train must be split to isolate the burning coach at a safe distance from the remainder of the train. This may involve splitting the train into three portions. This is just an example, and common sense at the time will indicate the correct actions to be taken.

At the terminus

At the end of the journey we will need to disconnect the locomotive and run round the train. The train is brought to a stand clear of the points so that the locomotive can move forward and reverse past the train. With the train stationary, the vacuum is completely destroyed and the ejector closed. The direction of the locomotive is now reversed. It will be necessary to place the steam brake lever to the 'off' position and push the centre lever in, securing the lever with the hook. With the steam brake off, the regulator is opened and the locomotive will move slightly towards the train, buffering up; the steam brake is then applied and the regulator closed.

The fireman should apply the handbrake, and the steam brake lever can again be pinned up on the hook. It can now be considered that the locomotive will not move. Some drivers will leave the steam brake on and the handbrake off while the fireman is between the locomotive and train, but this practice will blow the oil from the steam brake cup. Also, if the driver inadvertently pinned up the steam brake the engine could move with the fireman still uncoupling; the same could happen if the ejector was open, the brake was in the 'off' position and the vacuum pipes had been replaced in the wrong order. Using the steam brake and not applying the handbrake is quicker, but this is another area where drivers adopt different methods to carry out the same job. Both have their advantages and disadvantages- the main thing to consider is safety.

When the fireman returns to the footplate after carrying out his duties, the driver applies the steam brake and the fireman releases the handbrake. The driver selects the direction of travel and, if all is clear, releases the steam brake to run to the buffers or behind the signal in order to clear the points. There is no need to use large regulator openings for such a short distance. It is far better to use small openings and allow the locomotive to coast, and the regulator can always be opened again. This saves unnecessary wear on the brakes and tyres of the locomotive. If the rail is greasy the sanders can be started before the locomotive is moved.

When the road has been set, the driver makes the run-round movement to bring the locomotive to the other end of the train for the return Journey.

As we made our outward journey chimney-first, the return journey will be tender-first, so it is advisable to water down the coal in the tender to lay the dust. If the rail is greasy, as the locomotive approaches the train the sanders can be operated to ensure a good start from the station. An important thing to remember here is that after buffering up to the train the driver must not open the cylinder drain cocks while the fireman is in between the locomotive and the first coach.

The return journey is similar to the outward one. During the day it will be the driver's responsibility to maintain a sufficient level of water in the tender tank. When running tender-first, the tender should be full if possible, as the additional weight will make it ride better and more safely.

As this is only a demonstration journey it is not possible to go into a great deal of explanation as to the many differences between the operations of different private railways.

THE SAME JOURNEY WITH A FORMER GWR LOCOMOTIVE

After creating vacuum by the use of the ejector, the locomotive is driven in a similar way to the 'Black Five', although it is not expected that the speed of the light engine will be sufficient to enable the ejector to be closed; this creates the vacuum when the locomotive is standing or moving at a speed of less than 15mph. Larger GWR locomotives have a small and large ejector fitted.

On arrival at the coaching stock the guard will call the locomotive on to the train; with the vacuum brake applied, the fireman will apply the handbrake before going in between to couple up. Again it will be seen that applying and releasing the handbrake should be the responsibility of the fireman.

After coupling to the train, the driver carries out the vacuum test as described above before moving away to the station. On the way, if the opportunity exists, a brake test can be carried out.

On arrival, as with the previous example journey, the vacuum is destroyed, the handbrake applied and the cylinder drain cocks opened to prevent a build-up of steam in the cylinders.

Great Western locomotives are fitted with a displacement lubricator and, if the train is to stand at the station for some time, the lubricator can be closed on the main gallery until nearer departure to save the oil. When the main valve on the gallery is re opened it is a good opportunity to check the flow of oil through the sight feed glasses, making adjustments as required. The flow should be one drop of oil every 30 seconds; at that rate there should be sufficient oil in the reservoir to last for a day's work.

The departure time is now close, so the vacuum is created with the small ejector (if fitted) or the single ejector to the required level of 25 inches, and the brakes applied to release the handbrake. When the handbrake has been released the vacuum is re created, and the train is held stationary by partly destroying the vacuum on the train pipe side of the brake system.

When the 'right away' has been given by the guard, the driver starts the train in a similar method to the previous example. Again the locomotive will be worked lightly until the line speed has been reached. The vacuum level is maintained by a vacuum pump on the crosshead of the locomotive, so when the speed of the engine exceeds 15mph the ejector can be closed.

When coasting with a Great Western locomotive the regulator is closed, then partly moved back to open a jigger valve fitted to the faceplate of the locomotive; the regulator valve will, however, remain closed. The cut-off on the reverser must be set at 45 per cent, and this will keep oil fed to the cylinders and valves.

There are two methods of braking with this type of locomotive. The first is similar to the LMS method just described. On the approach to the stop the driver allows the locomotive to coast, but will first have opened the ejector to give him complete control of the brake as with the 'Black Five'.

The second method must only be attempted with practice and preferably under supervision, as it is completely different. When the regulator is closed, shutting off steam early, and the locomotive is allowed to coast, the line speed will fall to below 15mph, and this will stop the crosshead vacuum pump from maintaining the vacuum level, and speed will start to fall further as the loss of vacuum causes the brakes to be gradually applied, eventually bringing the train to a stand. If the train is thus likely to stop short, the ejector is opened to release the brake and the locomotive will continue to coast. With practice the driver should be able to stop in the correct place without the use of the driver's brake valve.

As before, the time at the station is a good opportunity to take stock of the boiler water level, steam pressure and condition of the fire. Assuming all is well, the vacuum can be created with the ejector before the 'right away' signal is received.

At the terminus the run-round procedure is as before. With the vacuum destroyed, the driver reverses the direction of the locomotive. There are two methods of detaching the vacuum-braked engine from the train. The first is to re-create the vacuum to 15 inches on the train pipe and reservoir side; this will release the brake on the engine, leaving it still applied on the train. When the regulator is opened the locomotive will move to ease the couplings; the vacuum brake is then immediately applied.

A safer method is for the fireman to go between the locomotive and train and disconnect the vacuum pipes, placing the locomotive-side pipe on the dummy coupling, then returning to the footplate. If the driver now creates the vacuum the locomotive can be moved to ease the couplings, and the fireman can apply the handbrake before returning between the locomotive and train.

After the locomotive has run round the train as before, the guard will call the engine back on to the stock. The driver buffers up and the fireman applies the handbrake before going in between to hook on the locomotive.

When a GWR locomotive not in steam is moved by being towed by another locomotive, it will be necessary to open the vacuum relief valves, which will prevent vacuum from being built up in the system by the action of the crosshead pump; this would cause vacuum to be created on the reservoir side, applying the vacuum brake, even if the locomotive has no steam.

The regulator lever of a Great Western locomotive in the closed position (left). When coasting the regulator handle should be partly opened to open the jigger valve (right) and the reverser set at about 45 per cent cut off; this will keep oil fed to the cylinders and valves. *R. G. Fox*

Working with the Great Western steam and vacuum brake

This type of brake (see also Chapter 6) is fitted to most of the small GWR locomotives such as the pannier tanks. As with the LMS type and other steam brakes, the steam brake operates on the engine and not the train, and the locomotive has no vacuum chamber as with the GWR vacuum brake. However, as with the other GWR types it has train pipe and reservoir sides.

The ejector on this type of locomotive is built into the brake valve, controlled by one common three-position lever. With the lever to the right the steam brake will be applied; in the central or vertical position the brake will be 'off' in the running position; to the left the ejector will be operating.

Let us assume that we are about to make a light engine movement to couple up to the coaching stock. It will be necessary to remove one of the vacuum pipes from the dummy coupling to prevent vacuum from being built up in the system. The movement of the locomotive will cause vacuum to be built up on the reservoir side by the action of the crosshead pump, and this will apply the steam brake.

With the steam brake lever moved to the right to apply the steam brake, the handbrake can be released. These locomotives are usually fitted with a pole-type reversing lever, released by a spring catch in the quadrant. Just as before, the direction of travel should be selected before the steam brake is released.

With the locomotive buffered up to the stock, the fireman applies the handbrake before coupling up. It is not actually necessary to apply the handbrake with this type of locomotive, but as stated before the driver must compare speed with safety. When the locomotive has been coupled up to the train, the vacuum is created by moving the lever to the left to open the ejector; this will release the steam brake, but the train will be kept stationary by the handbrake. If vacuum cannot be created a check must be made that the vacuum pipe is on the dummy coupling. As with other GWR locomotives, the required vacuum level is 25 inches. When the vacuum has been created a brake test is carried out as before.

Before moving away with the train the direction of travel is selected and vacuum created to the required level; the locomotive and train can be kept stationary by applying the steam and vacuum brake. Driving the train is similar to other locomotives. To move away place the brake lever to the left; this will open the ejector. On the road, when the line speed reaches 15mph the ejector can be closed off and the crosshead pump will then maintain the vacuum.

Braking the train is much the same as with standard Great Western locomotives except that the steam brake will work on the locomotive and the vacuum brake on the train. If the brake lever is placed to the right, air will be admitted to the train pipe and the vacuum brake will be applied; at the same time the ejector will be closed off. The movement of the lever will also apply the steam brake to the locomotive in much the same way as the LMS combined steam and vacuum brake.

When the locomotive and train reaches its destination the vacuum is destroyed from the train pipe and the steam brake applied. This will leave vacuum on the reservoir side of the system, which will keep the steam brake applied on the locomotive and will have to be destroyed by the use of the jigger valve on top of the vacuum reservoir in the cab before

Driving a loose-coupled freight train requires a technique quite different from that of a continuously braked, close-coupled passenger train. *Tony Bond FRPS*

the locomotive can move to ease the couplings.

As before, the fireman should apply the handbrake before uncoupling. When the locomotive has been uncoupled the locomotive-side train pipe connection will have to be left off the dummy coupling to allow the locomotive to run round the train.

WORKING LOOSE-COUPLED TRAINS

Some freight trains are partly or fully vacuum-fitted, so their braking is similar to that of a passenger train, except that there are no direct admission valves fitted on the train pipe, and air is admitted to the brake directly from the driver's brake valve. If the train is completely vacuum fitted the brake will operate on all the vehicles, which should be screw coupled and buffered up with no gaps between the buffers. However, a part fitted train will be left with gaps due to the three-link couplings.

The driving of a completely loose coupled train is different from a vacuum-fitted train, mainly because it is comprised of wagons with non adjustable three-link couplings; this means that there is a gap of about 6 inches between the buffers of each wagon when the train is working. This makes the train difficult to stop. When starting the train, this slack in the couplings must be taken up gently, otherwise a snatch could cause the train to be broken in two. In addition, with a loose-coupled train the driver's brakes work only on the engine and tender, so the braking has to be more precise to stop in the right place. The brake must be applied with consideration and not too abruptly, or the unbraked section of the train might hit the almost stationary locomotive and tender with sufficient force to throw the guard against the body of the brake van. The rebound could then cause the couplings to snap, leaving the train divided.

Before a freight train is moved, the guard should inform the driver of the composition of the train and the number of vehicles, braked and unbraked, and the fireman must place the correct headlamp code on the front of the engine (see Chapter 4); as before, a lamp must not be left between the locomotive and train.

The train is started in the following manner. The regulator is opened for about one revolution of the driving wheels, then closed; this will take up part of the slack in the couplings. The locomotive is then allowed to move slowly forward for a short distance before the regulator is re-opened for about another revolution of the wheels, then closed again.

Depending on the length of the train, this will have to be repeated until the brake van is seen to move; this will indicate that the train is complete. The fireman and guard should exchange signals when the train is under way to indicate that it is complete. If the rail is greasy the use of the sanders will assist in the starting and, indeed, stopping of the train. It must be remembered, however, that the sanders do not come on the instant that the valve is opened, but take time to blow out any water in the system brought about by steam condensing in the pipes after they were last used.

When driving the train the controls are used just as with a passenger train as regards the regulator and cut-off, etc., until the train requires to be brought to a stand. The first thing to remember is the weight of the train, and that the braking effect comes from the locomotive and tender only; the unbraked weight of the train may exceed the weight of the locomotive many times, so the train must be brought under control early. With this in mind the driver should shut off early and allow the train to coast; this will save steam, and wear on the brake blocks.

Braking should start by applying the brake for a few yards, then releasing it, to allow the train to buffer up to the locomotive. Depending on the number of wagons and the total length of the train, this may need to be repeated to slow the locomotive and bring the train together buffered up. Only experience will tell when this has happened, then the brake can be fully applied to bring the train to stand.

DISPOSAL

At the end of the day, when the locomotive is being disposed by the fireman, the driver must examine it for defects and report any faults to the fitters. If the defects are not correctly reported when the engine is in steam, they may be missed should it undergo a seven-day examination by the shed staff when it is cold. Parts must be clearly identified on the defect card so that the fault can be located; for example, the 'left-hand side' must be assumed to be looking towards the chimney. When referring to items such as springs or wheels they must be stated as 'leading' or 'trailing'; for example, a broken front left-hand spring would be stated as 'the leading left-hand driving wheel spring is broken'. Items on the footplate, for example the gauge frame, can be described as 'fireman's side' or 'driver's side', so that they cannot be mistaken by the fitting staff. Ejectors must be described clearly as 'large' or 'small', not just 'ejector'. Likewise, possible faults with injectors must be correctly identified as to which side, left or right.

All tools, oil bottles and lamps should be put back into the stores for use again. Before leaving the locomotive the gauge glasses should be turned off or isolated, and all stop cocks should be closed. The regulator must be closed and the locomotive should be in mid-gear with the handbrake hard on. The cylinder drain cocks should be left open; if the locomotive has a mechanical lubricator this will also close the steam supply to the atomisers. Any stop cocks outside on the framing of the locomotive should be closed.

It is always advisable to have a last look round to make sure that nothing has been missed before leaving the locomotive. As with the fireman, at the end of the day it is polite for the driver to thank his mate for his assistance during the day - after all, without him the locomotive would go nowhere. Indeed, the whole day's work could not take place without the teamwork of all the railway staff.

The end of the day

What nicer way to finish than over a pint or two with friends talking over the events of the day on the railway? Everyone has shared the common satisfaction of seeing trains operated and meeting people with the same interest.

The preserved operating railways of this country give great pleasure to the general public as well as to the enthusiast. It is nice to finish a shift knowing that you have done your best and brought a lot of enjoyment to many people, without whose patronage heritage lines would be difficult to maintain.

Glossary of techinal terms

Anti-vacuum (or snifting) valve
A valve fitted to the superheater header or the steamchest of a locomotive, which opens automatically when the locomotive is coasting to prevent the vacuum in the cylinders from drawing ash from the smokebox into the steamchest.

Articulated locomotive
A locomotive with a powered bogie at each end sharing a common firebox and boiler. There are several similar designs such as the Garratt, Double Fairlie, Mallet and Meyer, among others.

Ashpan
A container below the firegrate to collect the ash and cinders from the fire. It is also fitted with doors, known as dampers (qv), to control the air admitted to the firebed.

Atomiser
A device that emulsifies lubricating oil by mixing it with steam before it is fed to the cylinders.

Axlebox
The container of the bearing in which the end of an axle rotates. The axlebox moves vertically in guides known as horn guides (qv), the movement controlled by springs.

Baffle plate
A plate fitted in the firebox to deflect the cold air entering the firebox through the firedoors and prevent it from coming directly into contact with the tubeplate.

Balance weights
Weights fitted to the driving wheels of locomotives to offset the weight of the crank pins and coupling rods.

Big/little ends
The larger/smaller ends of a connecting rod.

Bissel truck
A truck that carries two wheels under the footplate of a locomotive.

Blastpipe
The pipe in the smokebox through which spent steam exhausts through the chimney to the atmosphere. It also converts the exhaust into a draught, drawing air through the firebed and hot gases through the boiler tubes.

Blower or jet
A steam-operated device fitted to the top of the blastpipe to create a partial vacuum in the smokebox when raising steam. It also helps to prevent a blow-back through the firedoors should the smokebox vacuum be destroyed.

Bogie
A truck with two axles placed before or after the driving wheels to help steer the locomotive and carry some weight.

Brick arch
An arch of refractory bricks fitted into the front of the firebox on the tubeplate, protecting the tubeplate from flame damage and assisting combustion of the gases in the firebox.

Clack valves
Self-operating non-return valves fitted to the injector allowing water to be admitted to the boiler against boiler pressure and preventing steam from being blown back out of the boiler.

Continuous blowdown valve *see* Scaling

Crown sheet
The part of the firebox inner wrapper plate that forms the top of the firebox.

Cylinder drain cocks
Cocks fitted to each end of a locomotive's cylinders and controlled by the driver. They drain any water or condensation from the cylinders to prevent the cylinder ends from being blown out by water pressure.

Dampers
One, two or, in some cases, three doors fitted to the ashpan to control the air supply to the fire.

Dead centre
The position of the crank pin when the piston changes direction in the cylinder.

Diaphragm
The piston in a vacuum chamber. Diaphragms are also used in brake assemblies to control the steam brake from the vacuum brake.

Displacement lubricator
A lubricator in which steam is condensed into hot water, which emulsifies with the oil and displaces it to lubricate the cylinders.

Dome
A raised section on top of the boiler barrel where the hottest and driest steam collects; it also sometimes contains the regulator valve.

Drag box
The location of the connection between the tender and locomotive drawbars.

Eccentric
A crank usually fitted to the driving axle and having a different centre from the driving wheels, used to drive valves, etc. Usually called a 'return crank'.

Ejector
A steam-powered device that expels air from the train pipe in a vacuum brake system.

Firebars
Cast iron bars fitted in the firebox forming the grate.

Flue tubes
Large-diameter boiler tubes containing superheater elements, fitted above the small smoke tubes.

Foundation ring
A ring of square or rectangular section fitted to the base of the firebox between the inner and outer firebox wrapper plates.

Fusible plug
One of several plugs fitted to the crown sheet of the firebox, the central core of which is made from lead so that should the boiler become low in water the lead will melt and relieve the pressure on the boiler and warn the crew.

Gauge glass
A pillar of water in a glass tube on the boiler faceplate indicating the water level in the boiler; there are usually two and they are each encased in a gauge frame to protect the crew should the glass shatter under pressure.

Gland
A steamtight packing made from cast iron or soft material fitted round such components as piston rods or valve spindles.

Hom guides
Polished steel guides fitted to the main frames of the locomotive or tender to guide the vertical movement of the axlebox.

Hom stay
A heavy steel bar usually fitted across the gap at the bottom of the horn guide to prevent the horn from spreading.

Injector
A device-powered by live or exhaust steam that forces water into the boiler.

Jet *see* Blower

Jigger valve
A valve used to release the vacuum from the reservoir side of the vacuum brake in order to release the steam brake.

Journal
The polished surface at each end of an axle that runs in the white metal of the axlebox bearing.

Lap
The amount by which the valve overlaps the steam ports when the valve is in mid-stroke.

Lead
The amount of steam admitted to the cylinder before the piston reaches the end of its stroke.

Lip plate
A cast iron half-ring that protects the lap joint under the firehole door and carries the baffle plate.

Loose-coupled
A term used to describe a train not fitted with a continuous brake.

Manifold
Sometimes called the 'steam stand', it is the junction from which steam is distributed to the injectors, carriage warming system, sanders, etc.

Motion
The arrangement of cranks, rods and levers operating a locomotive's valve gear.

Mudhole doors
Apertures fitted to the firebox and sometimes the boiler barrel to enable a boiler inspection to be made.

Pony truck
A leading pair of wheels that helps to steer the locomotive and carries some of the weight.

Pressure relief valves
Valves fitted to the cylinder end covers to release any water that may have built up in the cylinders. They usually open at about lOpsi above boiler pressure.

Primary air
The air admitted through the ashpan to the firebed (cf Secondary air).

Priming
Sometimes called 'hydraulicing' or 'carry over', this occurs when water from the boiler enters the main internal steam pipe and the cylinders.

Sanders
Steam-powered or gravity-operated devices that inject sand from a container fitted to the main frames on to the rails in front of or behind a locomotive's wheels to prevent them from slipping on greasy rails.

Saturated steam
Non-superheated steam (cf Superheated steam).

Scaling
Deposits caused by the salts in the water, which, if not removed by the continuous blowdown valve, would cause damage to the boiler.

Secondary air
Air entering the firebox via the firehole doors to aid combustion of gases in the combustion chamber (cf Primary air).

Sight feed lubricator
see Displacement lubricator

Snifting valve *see* Anti-vacuum valve

Stays and crown stays
Threaded metal rods fitted to the firebox to separate and secure the boiler plates round the firebox sides. They are riveted over on their outer ends and secured with nuts on their inner ends to protect them from flame damage. Crown stays are fitted to the firebox roof to secure the crown sheet (qv) to the outer wrapper plate.

Superheated steam
Steam dried and greatly heated in a superheater, which expands without condensing as rapidly and is more elastic.

Taps *see* Cylinder drain cocks

Thermic syphon
A pipe or tube fitted inside the firebox from the base of the tubeplate to the firebox crown to assist the convection currents in the water.

Tractive effort
A theoretical calculation giving an indication of a locomotive's ability to haul, or start, loads, rather than denoting its actual power. It is usually calculated by the following formula:

$$\frac{0.85\, d^2\, s\, n\, p}{2w}$$

where d = cylinder diameter in inches

s = piston stroke in inches

n = number of cylinders

p = boiler pressure in psi

w = driving wheel diameter in inches

The 0.85 represents 85% of boiler pressure, allowing for loss of pressure between boiler and cylinders.

The above calculation is for a two cylinder locomotive; for three cylinders multiply by 1.5, for four cylinders multiply by 2.

Underkeep
The bottom part of the axlebox, which usually carries the oil for lubricating the axle bearings.

Washing out
As the water boils in the boiler it gives off salts that form a sludge in the base of the water jacket. These salts need to be washed out otherwise they will cause bad steaming of the boiler (cf Scaling).

Index

Page numbers in **bold indicate main entries,** while *italics indicates an illustration.*